MW00720346

CHRISTY CLARK

CHRISTY CLARK

Behind the Smile

JUDI TYABJI

Victoria | Vancouver | Calgary

Heritage House Publishing Company Ltd.
heritagehouse.ca

LIBRARY AND ARCHIVES CANADA CATALOGUING IN PUBLICATION
Christy Clark: Behind the Smile – Judi Tyabji

Judi Tyabji, 1965-, author   Christy Clark : behind the smile / Judi Tyabji.

Includes index. Issued in print and electronic formats.
ISBN 978-1-77203-106-5 (bound).—ISBN 978-1-77203-107-2 (epub).—
ISBN 978-1-77203-108-9 (pdf)

    1. Clark, Christy. 2. Premiers (Canada)—British Columbia—Biography.
3. Women politicians—British Columbia—Biography. 4. British Columbia—
Politics and government—2001-. I. Title.

FC3830.1.C53W54 2016   971.1'05092   C2015-907561-0   C2015-907562-9

Interior book design by Setareh Ashrafologhalai
Jacket photo by Dave Chan

The interior of this book was produced on 100% post-consumer recycled paper, processed chlorine free and printed with vegetable-based inks.

We acknowledge the financial support of the Government of Canada through the Canada Book Fund and the Canada Council for the Arts, and the Province of British Columbia through the British Columbia Arts Council and the Book Publishing Tax Credit.

20 19 18 17 16   1 2 3 4 5

Printed in Canada

This book is dedicated to my father, Alan Tyabji, who always encouraged me to pursue my dreams, and to my stepmother, Barbara Tyabji, who keeps him smiling.

CONTENTS

If we strive to answer life's call with our best selves, we will be better people. Through life's challenges, we forge character.

PREMIER CHRISTY CLARK

It's not your obligation to complete the work,
but neither are you free to desist from beginning it.
THE JEWISH MISHNAH

INTRODUCTION
AND METHODOLOGY

APPROACHED RODGER TOUCHIE of Heritage House with the idea for
this book without ever consulting Premier Christy Clark, and I had a
contract in hand before she knew the idea existed. I had decided that
I would write the book whether she co-operated or not. If that sounds
insensitive, I should say that when I decided to write my first two books
about BC politics, I did not consult with either Liberal Leader Gordon
Wilson (who was, at the time, my fiancé) or former NDP Premier Glen
Clark, the subjects of those books.

About two weeks later, I had a chance to let Premier Clark know
that I had a contract to write a book about her. To my relief, she agreed
to co-operate.

She is exceedingly busy, and I was able to speak with her only once
during the writing of the first fifteen chapters. I want to emphasize that
Premier Clark's involvement has been arms-length at best and *Christy
Clark: Behind the Smile* is not an authorized biography. In fact, it's not
really a biography at all because she's still premier. This is a book about
Premier Clark written by someone who has known her, through politics,
since the mid-1980s. I am grateful to her for her bravery in granting me
interviews even when I told her I would not be providing her with a draft

copy of the book. Her trust that I would be fair in my writing is humbling. I have done my best to produce a book that is timely, relevant, and fair, and that justifies the trust the premier has placed in me.

You will read comments from many people I have interviewed. In conducting interviews, I usually first contacted the subject and spoke with him/her in person or on the phone. I emailed questions that they could use as a guide or ignore; the person being interviewed always had control over what we discussed and could opt to not answer a question. After our interview, I emailed my notes to the interviewees, and they were able to either make corrections or changes, and then the final version was sent back. In a few instances, entire interviews were conducted via email.

I am particularly grateful to Keith Baldrey and Vaughn Palmer for their generous input at the outset of my work, and for not laughing at my obvious ignorance of many political events that took place during the years when I was not engaged. My thanks as well to Sharon White, Brad Bennett, Grand Chief Stewart Phillip, Grand Chief Edward John, Chief Robert Louie, Chief Ellis Ross, Chief Ernie Crey, former Musqueam Chief Wendy John, former Tsawwassen Chief Kim Baird, Rich Coleman, Chip Wilson, John Reynolds, Moe Sihota, Tom Sigurdson, Dr. Pamela Cramond-Malkin, Don Guy, Laura Miller, Jas Johal, Mike McDonald, Jean Teillet, Linda Reid, Derek Raymaker, Dan Doyle, Floyd Sully, Bruce Clark, Mark Marissen, Lee Mackenzie, Dawn Clarke, Ed Henderson, Bob D'Eith, April White, Jason Down, Jenny Garden, Jatinder Rai, Athana Mentzelopoulos, Guy Monty, Josie Tyabji, and Gordon Wilson for providing time for interviews or input into this book. A special thank you to my stepson Mathew Wilson, who provided needed insight in the First Nations chapter, and my friend Monica Hassett, who put aside everything to provide essential help at the editing stage. A very special thank you to Erika Luebbe and the staff at the BC Legislative Library, who made it possible for me to research decades of news stories from home.

I should point out that one thing you will *not* find in this book are details about the premier's romantic life. I really struggled with whether to ask her the questions about whether she has or wants a boyfriend, and whether or not she plans to marry again. I knew if I asked her, she would answer. In the end, I could not imagine asking those types of questions of her if she was a man. I know that previous premiers were not asked about girlfriends, and so leaving her personal life alone seemed fair play.

Each of the chapters of this book begins with an excerpt from one of Premier Clark's favourite books, *The Road to Character*, by David Brooks. Her copy is dog-eared and full of notes. I asked Premier Clark to pick out her favourite passages from this book, to reveal to all of us why this resonated for her. I had to nag her to do it, since apparently she's busy running the province, but she did finally get it done in our second-last interview, while I sat and made small talk with her staff member. The wait was worth it.

Finally, since embarking on this project, I have often been asked to describe my relationship with Premier Clark. It took me some time to determine the best answer, which is that I consider her a colleague. She is also my premier, and therefore, when I am speaking with her, I address her as such unless I ask her permission to address her informally. Although I'm sure we could be friends, the truth is we have never spent any time together that is not directly connected to politics, and therefore, we are not friends or buddies, even if in pictures at some of these functions we look like "girlfriends."

Thank you to the Heritage House team, especially Rodger Touchie, for his detailed input and notes as I sent chunks of the book to him, and to Leslie Kenny and Lara Kordic for input before the edits. A wonderful shout-out to Tamara Davidson who juggles the premier's schedule. Now that is a magic trick. I appreciate that she was able to find times to slot me in for the interviews.

A very special thank you to my husband for his amazing support, for taking on many of my farm chores and work schedule, and for making sure I had time to conduct interviews and write this book when our family life had considerable unexpected challenges in the fall of 2015 and early 2016. We were both very busy, and he did not have time to read even a paragraph of this book before its publication, and in many ways, that is likely best.

Anger cannot win. It cannot even think clearly.

DWIGHT EISENHOWER[1]

CHAPTER 1

OM THE BOOK

IT WAS A big fight over yoga that led to this book about BC politics. Even the phrase *a big fight over yoga* seems uniquely British Columbian.

In the summer of 2015, the premier announced a public event called "Om the Bridge." It was to be an opportunity to participate in yoga on the historic and picturesque Burrard Street Bridge alongside Terry McBride of YYoga, a sponsor, and representatives from the Vancouver Yoga Association. Reports suggested that Chip Wilson, the founder of Lululemon, would also join the event and that Lululemon, the iconic Vancouver-based global yoga clothing company, was a sponsor. The announcement came on June 5. Om the Bridge was slated to run from 7:00 to 11:00 a.m. on the morning of Sunday June 21. Everyone was welcome.

It was billed as a "bridge to India," a country with which BC hoped to expand business ties in the wake of Indian prime minister Narendra Modi's visit two months earlier. Premier Clark's government invited people, using the hashtag #OmtheBridge. Clark wanted the event to be the largest yoga event on International Yoga Day outside India. Given the west coast's strong connection to yoga, and the premier's own personal use of yoga for fitness, it looked like a large event was a strong possibility.

4

BC was in the middle of a heat wave that had begun in early May, very unusual for the coast, and outdoor activities were increasingly popular because our weather patterns were locked into an extended summer. I must admit I was not paying much attention to the yoga event until comments about it started to fill up my social media news feeds, especially Facebook, and I started reading very angry comments directed at Premier Clark. They seemed so odd in connection with a free outdoor yoga event that I started to pay attention to try to understand what was going on.

It turned out that National Aboriginal Day, a day that had been growing in profile and popularity in BC over the previous ten years, also fell on June 21. It was also Father's Day, but the growing backlash stated the timing of the yoga event was a sign of disrespect to Aboriginal people, even though the yoga event was over at 11:00 a.m., and the provincial government's afternoon of Aboriginal Day festivities began at noon. There was something else playing into all this as well: the *Truth and Reconciliation Commission Report* was released by the Government of Canada on June 3, 2015. This report followed years of public hearings, research, and commentary from thousands of First Nations people and was the comprehensive report of decades of abuse and neglect. The executive summary alone was 536 pages and detailed many painful stories from Aboriginal people of all ages in their interactions with government and government programs, plus ninety-four recommendations for change.

This report represented a turning point in the dialogue with and about Indigenous people in Canada. Its release had a powerful impact across the country, and clearly the feelings of many Aboriginal people were raw. The government of Prime Minister Stephen Harper was not putting a lot of emphasis on the report, and many people wanted Aboriginal issues and stories given more prominence. For many, it was critical that the Truth and Reconciliation recommendations be given adequate profile.

The fact that the Truth and Reconciliation Commission was a federal matter and Om the Bridge was being hosted by the province didn't prevent the province from being the target. NDP Leader John Horgan put out a statement very critical of Premier Clark.

"I think that's offensive to people, particularly people who are struggling. People are trying to make ends meet and they see the highest priority of the leader of the government is to roll down mats, close a roadway on Father's Day, on National Aboriginal Day, and she's going to say 'come look at me. I like to do yoga.'"[2]

He then tweeted a photo of himself pretending to do yoga in a suit.

Another prominent British Columbian who entered the fray was beloved children's singer-songwriter and folk musician Raffi Cavoukian. Also known for his social media posts, especially on Twitter, and for his child advocacy and strong ecological opinions, Raffi comes across on social media as a strong, passionate advocate for the weak. He is also extremely political and outspoken in his views. His views on Premier Christy Clark emerged most prominently during the #Omthe-Bridge controversy when he took a lead role online speaking out against the premier's plans.[3] In *Metro News* on June 6, Raffi began using the hashtag #ShuntheBridge and described the event as a stunt sponsored by corporate energy company AltaGas, based out of Alberta. Raffi urged Vancouverites to stay home, saying that corporate sponsors of a yoga event on Aboriginal Day means "something stinks."[4]

Other critics said doing yoga on an oil-soaked bridge in the blazing sun was not a good idea. The premier was not deterred.

"I don't think there will be a picture in the world more beautiful than one of thousands of yoga mats on the Burrard Street Bridge on, hopefully, a sunny day."

The negativity continued to grow: the bridge closure was inconvenient; the cost of the event was $150,000; the corporate sponsors, it was said, had also donated to the BC Liberal Party; and increasingly, the event was said to be disrespectful of Aboriginal people.

Terry McBride of YYoga claimed surprise over the public criticism of the event and bridge closure. He told CBC he felt it was because people didn't have the facts, including that the bridge closure would only be from 7:00 to 11:00 a.m. on a Sunday. When asked whether International Yoga Day celebrations would detract from National Aboriginal Day celebrations, he sounded somewhat mystified.

"I don't know why one can't celebrate two great things at the same time."

Meanwhile, Christine Brett from the Nuxalk First Nation told CBC that protests were planned at the Burrard Street Bridge and in Victoria.

"I just find it a slap in the face that [Premier Clark] doesn't even recognize National Aboriginal Day."

On June 11, six days after the event was announced, Vancouver Mayor Gregor Robertson announced that he would not be attending Om the Bridge and would attend National Aboriginal Day celebrations instead.[5]

That same day, Premier Christy Clark tweeted a photo of herself that seemed to bring the rage of the opponents to a head. In the photo, she was standing in front of a Taoist Tai Chi dojo with the words "Hey yoga haters—bet you can't wait for international Tai Chi day." Social media blew up with responses, her opponents attacking her for missing the point, her supporters retweeting her post and arguing for a sense of humour.

Online, things had begun to spiral out of control. I put out an editorial on my Facebook page, expressing confusion about the anger and hoping someone could explain the extreme response, but the comments didn't help me understand. From my small sheep farm on the coast of BC, I felt I had an interesting perspective on the whole controversy. At the time, I had over four thousand friends on Facebook from political and non-political backgrounds, and this gave me a wide range of opinions floating past on my news feed. Almost all of my NDP friends were posting and sharing comments against the event similar to Horgan's. Many of my non-political Aboriginal friends seemed persuaded that the premier was insulting them with the event timing. I received a few private messages from Aboriginal friends who had planned to attend the event and were confused by the negativity, asking me if they should cancel their plans.

Soon the hashtag #OmtheBridge turned into #ShuntheBridge and then #DrumtheBridge, which apparently called on people to show up and make noise to disturb the yoga.

The overwhelming emotion I was seeing was rage, tinged with hatred, and it was feeding on itself. On the morning of June 12, things reached a boiling point in the form of an ugly Facebook event with over 980 confirmed attendees. It was called FUCKING BURRARD BRIDGE HARSH NOISE.

The description summed up spirit of the thing pretty succinctly:

WE'VE GOT GENERATORS! WE'VE GOT P.A.S! WE'VE GOT SCRAP METAL! BRING YOUR HARSH NOISE AND RUIN THIS HIPPY HORSE-SHIT! FUCK YOU CHRISTY CLARK AND CHIP WILSON, UNDER THE FUCKING PANZER TREADS! FUCK YOU!

The agenda for this event was more disturbing. It suggested vandalism and violence, including:

1 PM—RAISING THE SEVERED HEAD OF CHRISTY CLARK ON A PIKE

2 PM—LOTTERY FOR WHO GET TO SKIN CHIP WILSON ALIVE

I saved a copy of the event page and called the Vancouver Police to report a hate crime against the premier and Chip Wilson. Then I reported the page to Facebook as a hate crime with a credible threat of violence, and the page was down within the hour.

After a week of escalating controversy, the premier stated she would not attend and then that the event was cancelled.

"Unfortunately," said the premier, "the focus for the proposed Burrard Street Bridge event has drifted toward politics, getting in the way of the spirit of community and inner reflection."[6]

I was relieved that the event had been cancelled, but I was upset about the entire conversation and its baffling escalation. I was not angry that people were angry; people have a right to be. I was upset that people felt comfortable setting up an event to murder the premier and a prominent member of the business community, and that almost a thousand people had agreed to attend it. If that site had been about then–prime minister Harper, I'm sure that the perpetrators would have been in jail the same day.

It bothered me, at a fundamental level, that the public dialogue had provided consent for such unbridled hatred. That the level of hatred was directed at someone who had set up a free family event to do yoga in celebration of international peace seemed ironic and surreal.

Jas Johal is a former Global BC TV reporter, a well-known Indo-Canadian, and the current director of communications for the BC LNG Alliance. "The controversy seemed to be promoted by non-Asian, city-focused activists. They overlooked that this was a promise made to Indian Prime Minister Modi, when he was visiting, and that it was all part of the attempts to build a stronger bridge to India. There should have been some push back on the messaging."

As an immigrant from India myself, I agreed. I had been happy to see the provincial government seeking stronger ties to India, and a community celebration including yoga seemed perfect.

Multicultural strategist Jatinder Rai had concerns about the controversy."[The premier] felt the yoga event was an opportunity to put this beautiful province on display so people could see it worldwide and be part of an international event. People ask, 'How did you pick the day?' and she didn't pick it, it was picked by the UN. People criticized her for holding it the same day as National Aboriginal Day, but she didn't pick it. There were many other yoga events all over BC, on the same day as Aboriginal Day, but no one complained about that."

But this, of course, misses the larger question. It's not really about what people felt but rather *why* they felt it.

The controversy had not only occurred on Facebook. At the height of the furor, *Tyee* contributing writer Bob Mackin raged over insider sponsorship from Lululemon, taking to social media with what he considered evidence that Lululemon, as a sponsor, was also registered to lobby the premier. He listed the donations by Lululemon founder Chip Wilson and the company as if some terrible conspiracy were afoot. His tweets were referenced by online bloggers, and the frenzy escalated.

When the major media ran a big headline that the event had been cancelled after the sponsors had pulled out, Lululemon was listed most prominently, but according to Chip Wilson, Lululemon had never agreed to sponsor the event at all.

Wilson first met Christy Clark when she was hosting her radio show at CKNW in Vancouver. "I got to know a person who was incredibly flamboyant, inquisitive, and smart. It was a wonderful interview and I really enjoyed meeting her." As for Om the Bridge, "The misconception is that I sponsored it. Really, the premier in her visioning about this event, that would be great for BC, was that on Yoga Day we would have the Burrard Street Bridge that was full of people doing yoga, and the sun coming up and a helicopter shooting a picture, with the Georgia Strait and the north shore mountains, and this image would go around the world, with all the beauty of Vancouver. Who could be opposed to that?"[7]

Wilson loved the idea. However, the difficulty for him was that he was no longer in a decision-making position at Lululemon. After the premier explained her vision, her team took over, and Wilson opened some doors to the new management team; however, "They turned it down as a sponsored event. I was surprised by that because if the premier wants something, you should say yes."

Lululemon didn't pull out of the event because, in fact, they had never been part of it. All the allegations that Mackin and his blogging fans made regarding Lululemon trying to cozy up to the premier, or that the premier was corrupted by political donations, were off-base.

Sadly, that is typical of politics in BC.

British Columbia has its own distinct political culture. People from other jurisdictions often have a hard time understanding it, and frankly even within BC, many people struggle to navigate the complicated dynamic of our amazing province.

This is my third book about BC politics, and I find myself compelled to write it in much the same way I felt compelled to write *Political Affairs* (1994) and *Daggers Unsheathed* (2002). Here, I hope to map out some of the backstory of British Columbia's complex political landscape and peel away a few layers, so that the reader has a bit of context to understand the unique leadership of Premier Christy Clark.

Much of our character talk today is individualistic,
like all our talk, but character is formed in community.

DAVID BROOKS

CHAPTER 2

YOUNG LIBERALS AND TRUE BELIEVERS

FIRST MET CHRISTY Clark in 1984 or 1985, while studying at the
University of Victoria. Christy and I were both Young Liberals and
met at a few functions in Vancouver and Burnaby. At that time she
was attending Simon Fraser University in Burnaby, so we didn't know
each other that well. The gatherings were usually well attended, with
anywhere from thirty to eighty youth, depending on whether it was an
informal gathering or a function of the Liberal Party. Christy Clark and
I were nineteen or twenty at the time and both of us "true believers." My
parents were Liberals, and I'd been recruited to the Young Liberals on
campus. Christy was already involved because of her father's activities.
When I think back to those first few meetings, what I remember most
about her was her laugh, her curviness, and her hair. She was always
smiling and interested in the latest political conversation.

Before I get into what I mean by true believer, perhaps a little background on the political realities of the day. At that time, the national
Liberal Party and the Liberal Party of BC were the same political party,
which is not the case anymore. The Young Liberals (YLs) was a fairly
active club, even in BC, although provincially, the Liberals had been in
the wilderness for a long time.

In Canada, federally, we generally have three national political parties: the NDP, considered to be left-wing; the Liberals, considered to be centrist; and the Conservatives (including the Reform Party), considered to be right-wing. The leadership of these political parties will usually determine exactly where on this political spectrum a party will stand, and membership of the parties is usually somewhat flexible in supporting the leader's positions on this spectrum.

In the Canadian federal election of 1984, Brian Mulroney and the Progressive Conservatives defeated John Turner's Liberals and formed a majority government, the first stable Conservative government since the late fifties/early sixties, Joe Clark having lasted less than a year before falling to a non-confidence vote in 1980.

In British Columbia, the political landscape was, and is, unique. When I first met Christy Clark, the BC legislature was divided between the Social Credit Party and the NDP. Social Credit came to power in the 1950s as a result of a special ballot set up to try to keep the labour-affiliated party out of power. This ballot elected the Social Credit Party, which evolved into a coalition of the provincial Liberal and Conservative parties, described in detail later.

Social Credit became almost unbeatable, forming ten of the eleven governments from 1952 until 1991. What this meant was if you were a Liberal in the mid-1980s, you supported the Liberals federally and Social Credit or NDP provincially, unless you were a so-called true believer.

So what is a true believer? There are different definitions, but when it comes to politics, a true believer is a person who holds to certain core political beliefs even when these beliefs do not help gain power. In the mid-eighties, anyone who was prepared to volunteer time or contribute money to the Liberal Party of BC, hoping to elect people, was called a true believer. I first heard the term directed at me shortly after I joined the Young Liberals at eighteen. It was used mockingly by a fellow YL who supported Social Credit, who, I'm sure, thought of me as a bit of a dweeb.

In politics, the true believers are actually gold during tough times, because they stick around when a party is out of power and has no money, because they believe in what the party stands for. Most important, they believe their political party would make the world a better place. True believers will work for no money and no expectation of immediate reward. They will slog away between elections, work like slaves during elections, run for office, contribute money, and argue their party and candidate's positions with the passion of their beliefs and conviction.

True believers make up the backbone of the political system because these are the people who do most of the work. A political party *can* run without them, if it has a lot of money, but even money can't always buy power. Have you ever been called during an election? If you are talking to a volunteer who seems genuinely passionate about a candidate or party, as opposed to a telemarketer with a script, that person is likely a true believer.

However, in the 1980s in BC, Liberal true believers were also seen as enemies of the Social Credit Party, because every vote or dollar we drew into the remnants of the provincial wing of the Liberal Party was seen as a vote or dollar drawn away from Social Credit, thus weakening it and leaving it more vulnerable to the NDP.

Christy Clark was a true believer both in the Liberal Party of Canada and the Liberal Party of BC. In fact, she was a second-generation true believer, following in the footsteps of her father. She was passionate about Liberal issues like social justice, multiculturalism, equality of opportunity, fiscal responsibility, economic development, democracy, health and education, and the environment; she was known as a progressive thinker.

There are often arguments about what it means to be Liberal in Canada, although there is usually agreement that Liberals support social programs (such as health care and pension plans); targeted government spending on social and government infrastructure (such as post-secondary education or national highways); tax policy that encourages economic growth or business investment; scientific work tied to economic development, international policy that encourages immigration and multiculturalism; and fiscal responsibility, including balanced budgets.

By way of contrast, the NDP generally supports higher taxes on business and high-income earners to support expanded government programs; labour laws and policy that provide support and higher wages for unionized workers or minimum-wage workers; laws and regulations that protect the environment; government social programs that support families (such as public education and daycare), government-supported affordable housing projects; and deficit spending to cover the costs of the expanded government programs. Conservatives tend to support smaller government, lower taxes (especially on business), reduction in government spending and social programs, and restricted immigration.

True believers within a political party tend to be quite passionate about their party's position on key topics. True-believer Liberals in Canada and British Columbia feel passionately that a tolerant, multi-cultural mosaic enriches Canada; that respect for Indigenous people must be woven into public policy; that every person deserves an equal opportunity to succeed; and that government's role is to provide the environment in which this opportunity can occur. What this means, specifically, is sound education, good health care, a healthy framework for a sustainable economy, and affordable government. Liberals will sometimes take on unions just as they will sometimes take on big business if either group appears to be acting contrary to the public interest.

I often tell my children that the world is run by the people who show up, because the truth is that most people *don't* show up, especially for politics. People who show up in politics are in two groups: those seeking power, influence, or a career, and those who want to help their political party make the world a better place, whether nationally, provincially, or even municipally.

This brings us back to the Liberal Party in BC in the 1980s, a party made up *entirely* of true believers. Often we would attend receptions alongside people who were pretty angry with us for believing in something that wasn't good for the Social Credit government. There were constant efforts to undermine any progress we made, and since many Socreds were federal Liberals, any time we raised money, they made sure it went to Ottawa and was not available for our efforts.

When I moved to Vancouver and started working for the federal Liberal Party in 1987, I attended a few Young Liberal social get-togethers, often held at someone's rented house or apartment. Often Christy was there, sitting in a chair, a bottle of beer in hand and looking entirely at ease.

I remember Christy Clark at these events as a young woman in a T-shirt and jeans and with long, wavy blonde hair. All the people were drawn to chat with her, and she engaged in conversations with a big smile on her face. Every now and then, her laugh would fill up the room and everyone would feel it. There were many people who attended, but few that I remember that well.

She was often among the smokers, so I didn't hang around too closely, because I'm allergic. In those days smokers smoked everywhere and it was part of the social scene, and I certainly noticed that she was popular and ready to mix it up when people wanted to debate.

As for me—the newly arrived, policy-wonk ethnic girl from the Interior—I was a non-drinker and very shy, usually overdressed and keen to debate the constitution. I didn't hang out with Christy, although we talked a few times and I really liked her. Everyone liked her. She reminded me of a sexy version of Peppermint Patty from the Charlie Brown cartoons, because she was at home with the boys, had a husky voice and a wry sense of humour. Her personality was warm and she seemed kind.

It is worth pointing out that, when I knew her back then, she didn't show any leadership ambitions whatsoever. To put that in context, there were many Young Liberals who would tell us, whether we wanted to know or not, of their ambitions to be a member of parliament or prime minister or member of the legislative assembly or premier. These were the YLs who would really focus on grabbing key positions or attention at meetings. As much as Christy worked at events, she didn't shoulder her way forward to grab key spots or seek out the limelight. She showed a keen intelligence and a gift for making friends, something not always easy for true believers willing to speak their mind.

Meanwhile, the split among Young Liberals between the young Socreds and provincial Liberals was pretty intense. Big arguments would break out. Some Young Liberals were actively engaged in organizing for the Social Credit Party, and many of these were loudly contemptuous of those of us hoping to elect Liberals to the BC legislature again.

Christy Clark followed every YL debate or discussion, and it was clear where her values lay, but her values seemed to serve more as a compass than an engine, guiding her opinions rather than driving them. Clark was always speaking up for the little guy, arguing against anything that she saw as elitist. When it came to the Constitution of Canada, she was ready to get into a scrap to fight for BC's place in Confederation and did not accept the view often voiced from Ottawa, namely that BC should not make waves and be happy with the status quo. Clark, like many BC Young Liberals, was pretty passionate about BC receiving its fair share in terms of Senate seats and federal ridings. In those days, we were seriously underrepresented in the House of Commons and the Senate.

Clark seemed comfortable with other people holding strong opinions and was not threatened by them, and she only became argumentative if she felt the opinions strayed from Liberal values. As true believers, we were pretty passionate about the need to be fiscally responsible and aim for balanced budgets, but this was within the context of ensuring that education and health care were properly funded. There was a general impression that, if government waste were to be addressed, it would allow for proper investment in the social programs.

She was a Liberal, provincially and federally, and her centrist perspective seemed woven into the fabric of who she was. That being said, I remember her as much more tolerant of dissenting views than I was; I was happy to argue with the Socred supporters in the Young Liberals, sometimes just to taunt them.

A look at the history between the Liberal Party of BC and the Social Credit Party helps understand what happens later.

IN THE 1952 election, the BC government was a Liberal–Conservative coalition. Keen to keep power, they introduced an Alternate Vote system to obstruct the success of the Cooperative Commonwealth Federation (later the NDP). In this system of voting, voters picked their first *and* second choice for government. The assumption was that Liberal voters would choose Conservatives second, that Conservatives would choose Liberals second, and that CCF voters would have to choose either Liberals or Conservatives second, meaning that no matter what, the coalition would have the most votes after the election.

It is a voting system that has not been used since.[1]

In an outcome that seems typically British Columbian: enough Liberals could not vote Conservative (even as a second choice), enough Conservatives couldn't vote Liberal, and enough CCF voters couldn't vote for either, that a *fourth* party, the Social Credit Party, took the lion's share of the alternate vote. In a truly "made in BC" outcome, this gave the Social Credit Party (known popularly as the Socreds) the most seats in the legislature, left the CCF in Opposition, and reduced the Liberals and Conservatives to "rump" parties.

A lovely bit of karma in that outcome, isn't there?

This began the Social Credit Party's dominance, and, soon after, the dynasty of the Bennett family of Kelowna, British Columbia. William Andrew Cecil Bennett (more commonly known as W.A.C. or "Wacky" Bennett) was a larger-than-life figure who, with the help of some colourful cabinet ministers, ran British Columbia as premier for twenty years.

The election of 1972 brought Dave Barrett's NDP into power for the first time in BC and reduced the ailing Liberal vote to 16 percent, winning David Anderson's Liberals only five seats. The NDP win shocked the political establishment, motivating anti-left-wing politicians to desperate measures, and helped W.A.C. Bennett recruit Liberal MLAs Pat McGeer, Garde Gardom, and Allan Williams to the Social Credit Party, an act that devastated the provincial Liberal Party.

Barrett's short reign left a legacy including the Agricultural Land Reserve, the Insurance Corporation of BC, and the institution of Question Period in the legislature. However, in 1975, the Social Credit Party returned to government, this time under the leadership of Bill Bennett, the son of W.A.C. In the 1975 election, the new Liberal Party leader,

Gordon Gibson, was the only Liberal elected, and the Liberals garnered only 7 percent of the popular vote. In the next three elections, voter support fell even lower.

Bill Bennett's Socreds won again in 1979 and 1983 before passing the torch to Bill Vander Zalm in 1986. In these three elections, not a single BC Liberal was elected to the legislature. From Gordon Gibson, the leadership of the Liberal Party of BC passed to Jev Tothill, then Shirley McCloughlin, and then Art Lee,[2] and though none of them managed to elect any MLAs, they all kept the party active and hopeful.

Mike McDonald was a seventeen-year-old Young Liberal when he first met Christy Clark at an SFU Young Liberal meeting in the fall of 1986. He describes himself back then as a geeky kid who didn't know what to expect from a political meeting.

"I remember Christy having the qualities we have seen in her today: full of life, approachable, lots of laughter. She made quite an impression on me, and we became fast friends.

"We quickly discovered that we both came from families of Liberals in dark times. The notion of struggle was intrinsic. Her dad, Jim, had run as a provincial Liberal in the late sixties and early seventies in Burnaby, outside the Liberal strongholds. He was a sacrificial lamb. My dad, Peter, ran in Dewdney in 1969 as a provincial Liberal, finishing third. They both fought the good fight, and both Christy and I were shaped by the experiences of our fathers: instinctively Liberal, not afraid of long odds."

They were soon tossed into the deep end with politics. "Within the first month of belonging to the SFU Liberal Club, there was a provincial election where Art Lee was party leader. He was respected, and we felt we could get behind him. There were many events and activities, and we both volunteered a lot. Christy's brother Bruce was on and off the ballot. I was very active in Maple Ridge. It was exciting to be part of something on the ground floor." The results of all that hard work were not great. "We came out of the provincial election with less than 7 percent of the vote. Alas, there were no Liberal seats, but our fervour was undiminished."[3]

It was an exciting time to be a Young Liberal in BC. Many of us attended meetings and helped with whatever volunteer jobs there were at the time. In those days, the federal and provincial parties were in the same office, and I had a job working for the federal Liberal Party in the fundraising wing, under John Turner's leadership. This meant I was woven into all the discussions that went on around the office.

At that time, the Liberal Party was also my social life. There were many teenage and twenty-something members, and we all felt very important to the party because we actually *were* very important; there was almost no one else around, because the party had no money. All the successful political folks were either in Ottawa with the federal Liberals, or with Social Credit in BC. There just were not enough Liberal members to do all the work, and many of us were thrown into key positions simply because someone had to do it, and we would do it for free. Sometimes experienced members trained us, and sometimes we just made it up.

Many of us, as young political geeks, ended up spending time at University Model Parliament (UMP), which meant that we participated in elections on campus to determine which political party would be government, which would be Opposition, etc. Then we invaded the real BC legislature in Victoria, and spent an entire weekend taking ourselves very seriously. We were so sure of the solutions and so passionate! We would work hard on legislation, negotiate the business of the debates, and engage in heated discussions with our colleagues privately and the opposing parties publicly. It was a lot of fun and great training for any young person wanting a chance to learn about government.

My cousin Mark Devereux and I attended the University of Victoria together, and we attended UMP before I graduated in 1986. In 1987 he attended when the Liberal Party was government. At that session, there were twelve Young Liberals in cabinet, and Christina (Christy) Clark was one of them. The elected leader of the Liberals was Patrick Salinger, and he appointed Christy Clark as Minister of Environment. My cousin Mark, well travelled by that age, was the Minister of External Affairs.

Mark admits that he remembers almost nothing about Christy Clark from that session. "There was a big head of blonde hair nearby, and I remember having some respect for everyone in the caucus; however, there were no indications whatsoever of future leadership ambitions emanating from the big bushy blonde head near me."[4]

Mike McDonald was another of the twelve Liberal cabinet ministers alongside Christy Clark in that 1987 model parliament and served as President of the Treasury Board. He remembers being awestruck at age eighteen by his surroundings on the floor of the legislature. He also remembers the NDP as an Opposition party. What is interesting is that in the 1987 list of twenty-six New Democrat attendees, listed among the four "additional members" was the name Adrian Dix.

Mike says model parliament was likely excellent training for Christy Clark, and he believes the leadership style she followed was obvious by the time they were in University Model Parliament a couple of years later, when she was Liberal leader and he was Liberal house leader. He tells a story about the last day, when he had to leave early because his duties on the Liberal Provincial Executive called him back to the BC mainland. As house leader, he was responsible for negotiating the deals with the NDP house leader and setting out the schedules for the laws that would be passed before they finished the session.

"I left on the bus, went to Vancouver, and was thinking how I finished all my work and cut all my deals. I was feeling pretty good about getting it all done before I left. That night I got a phone call [from Christy], and she told me that they had ripped up all my deals. She said the NDP went crazy, and it was a huge drama."

This was in the years before cell phones and the Internet, so decision making sometimes had to happen without consultation or communication because there were no mechanisms for the conversations to occur within the time frame in which the decision had to be made.

"Nothing is perfect," says Mike, "but you have to defend the party line and support the leader in order to succeed. You can fight decisions out internally, but once a decision is made, you have to accept it."

In the late eighties, we were plunged into constitutional discussions, which spread across Canada and led to some very heated debates. There were talks of Quebec separatism and western alienation, although few of the politicians in central Canada took the west too seriously. The federal

Liberal Party was humming; after all, we were the party that had been in power when so many dramatic events had taken place regarding the constitution. There was almost a feeling that it was the Liberal Party's responsibility to solve the problem of Quebec's dissatisfaction.

Meanwhile, it was hard to be heard in the west. We didn't have enough MPs to matter, and the election of the federal government was often over before our polls had even closed on election night. Still, discussions were heated, and we Young Liberals were often engaged in earnest debates about huge decisions—the kind over which we had absolutely no power whatsoever.

Christy's older brother Bruce was also politically involved. I remember he was often around at Young Liberal meetings. He was the tall, good-looking guy that all the girls were after. He had that air of understated competence that some people carry, a bit old-school, like Gregory Peck in *To Kill a Mockingbird*. He would attend meetings and then rush off to something obviously important. Christy and Bruce seemed very different and yet similar, as siblings often are—similar in looks and core values and dedication to the Liberal cause, but very different in personality and methodology. He seemed quiet and focused and ambitious and serious; she seemed relaxed, outgoing, fun-loving, and ready for whatever would happen next.

Bruce Clark explains where his little sister Christy engaged on the political spectrum. "I was just getting into the Young Liberals in 1975, and when Christy became involved, about 1980 or so, it was obvious that she was very good at it. She was on the more progressive side of the party; I was on the less progressive side."[5]

Being a progressive member of the party in the mid-1980s sets the stage for Christy Clark's working relationship with the leader of the Liberal Party who replaced Art Lee, in late 1987: Gordon Wilson.

That person . . . whose mind is quiet through consistency and self-control, who finds contentment in himself, who neither breaks down in adversity or crumbles in fright, nor burns with any thirsty need nor dissolves into wild and futile excitement, that person is the wise one we are seeking, and that person is happy.

CICERO

CHAPTER 3

CHRISTINA JOAN CLARK

C HRISTINA JOAN CLARK was born on October 29, 1965, the youngest of four children of Jim and Mavis Clark of Burnaby. Her older siblings are John, Bruce, and Jennifer, and there is a six-year age difference from oldest to youngest. In every conversation with people who knew the Clark family, it is clear that Christy shared a deep love with her parents, had a lively and supportive relationship growing up with her three siblings, and was surrounded by stimulating ideas and laughter throughout her childhood. Her parents were likely her strongest role models. It seems they opened the door to a world of possibilities to their baby daughter, and she ran through it.

Christy's father was a teacher with a passion for politics and the Liberal Party. Her mother held a master's degree, worked as a trained dietician, then a daycare operator, and later a family therapist who helped set up the Burnaby Family Life Institute to support families with challenges. Both of their careers allowed flexibility, which afforded the Clark family a lot of time together in their cabin on the Gulf Islands every summer, and on many holidays.

"One of the fantastic benefits of being a teacher," says Clark, "is all the time you get to spend with your kids, uninterrupted, when they are off school and you are off, too. We were able to spend two months of uninterrupted time up at the cabin, with Dad showing us how to chop wood, ride bikes, and go canoeing during the summer."[1]

Her parents shared a love of the creative arts. Her father expressed that through silk screening, which he used for cards and art, and of course for making political signs. Her mother was a painter, and some of her artwork is on the walls of Christy Clark's home in Vancouver.

The multicultural urban centre of Burnaby provided the city culture that shaped Christy Clark's world view, while the rustic natural setting of the family cabin grounded her in an awareness of nature and the ocean. It was a classic BC childhood, the combination of nature and city.

Clark's time at the family's Gulf Island cabin, from her early visits as a little girl to the trips she made later with her son, revealed changes to the environment that Clark would not have seen otherwise.

"Growing up outside for two months of the year, every summer, gave me an appreciation for the changes in nature. I have seen the anemones disappear from the tide pools, and as the ocean has warmed, I've seen the arrival of the red jellyfish. Climate change is obvious when you are able to be close to nature and observe it; the signs are all around us."

Clark believes that her siblings would identify her as the favourite child of both parents, stemming from the fact that she was the baby.

Bruce Clark certainly agrees with that. "Little sisters always get their own way."

He says that their parents created a home that was a rich environment for all their children, and that there were a lot of political discussions around their house. Jim Clark ran three times for the Liberals—in 1966, in a 1969 by-election, and in 1975—and Bruce thinks the home dynamic had a huge influence on how Christy interacted in the political world.

"She's my little sister. She has a strong personality. My mother referred to her as *jubilant*. She is positive, optimistic, and determined. When you are sitting at the dining room table and you are the littlest, and your dad is a political activist, heavily involved in politics and current events, in order to hold your own, you have to be able to punch above your weight. She was always able to do that."[2]

Christy says her mother was an enormous influence on how she perceived the world. "My mother was a gentle soul who was a helper. She was smart, introverted, strong, and felt things very deeply. She was capable of being firm. She didn't have any space in her life for negative words or energy. She didn't judge other people, ever, and she was very considerate of other people, always. Kindness was her first response. She had a talent for remembering to fill her life with things that made her happy, and jettison the things that, if you dwell on them, bring you down. She loved nature, and I learned my appreciation for the wilderness and nature from her, from the ancient forests to swimming in the ocean, the way she would revel in the beauty around us. A walk in the woods with my mom would take a long time because she would see so much."

It is through her mother's memory that Christy has a clear sense of herself as a child. "How people described me when I was young, and my mother said this to me first, is that I would be standing amongst a group of people, entertaining them, and engaging them. It was a busy house back in the sixties and seventies. There were a bunch of families around, and people had lots of kids. My brother Bruce would say I was always talking and trying to get everybody's attention. Typical youngest child behaviour. My sister and one of my brothers are quite reserved, while my eldest brother is loud."

Her parents were a huge influence on her interests, and she speaks of them with respect and admiration. "Both my mom and dad had a deep need to help other people. Interestingly, in both of their cases, they wanted to help other people live in healthier families and support children to fulfill their potential. In my dad's case, as a schoolteacher and counsellor, going above and beyond the daily schedule. In my mom's case, she started the first non-profit daycare in Burnaby at our church St. John the Divine. Later, with our parish priest, she co-founded the Burnaby Family Life Institute, which is still operating today." The pride in Clark's voice when she talks about her parents is obvious.

Christy Clark met Dawn Clarke, the woman she today describes as her best friend, when they were in grade two. Says Dawn, "In elementary school, we were mostly in opposite classes. What I liked about her right away was the huge smile she always had on her face. In those days, she would walk to school with another girl who lived near her, and they both had the same style of lunch kit, but with different characters. Christy's was a Holly Hobby lunch kit, and I just loved it. I was envious, and at the same time, I couldn't help but like her because she just had that personality. I remember thinking she was probably from a rich family because *wow*, that was some lunch kit!" She pauses, then adds, "I think she was wearing a poncho, which was very much in style."[3]

Dawn Clarke has a bubbly personality that projects through her voice as she tells stories from their childhood. She talks of going over to the Clark home for a birthday party shortly after she met Christy, and she says that there was an immediate sense of a lively, welcoming family. "Her mom had this beautiful demeanour about her, she was so nurturing and smart. She would make you feel instantly at ease."

Dawn says this was such a different environment than what she was used to at home, where the parents were in charge and it was pretty autocratic. In Christy Clark's household, you were encouraged to express your opinion.

Dawn's strongest memory of Christy in elementary school certainly foreshadows the future premier. "There was a contest among us kids, I think in grade two, to see who could hold their breath longest. She ended up passing out! She fell under the desk a little bit, but she came to right away. But she just had to win, didn't she?" Dawn laughs at the thought. "It's funny no one even alerted the teacher to this game. Well, she was fine. So obviously, her competitive streak was somewhere there all along."

Then in grade three or four, there was another glimpse of Christy Clark's sense of humour and drive. Dawn remembers classmates putting on some skits for the drama class that involved dance and music.

"We did one to 'Leader of the Pack' by the Shangri-Las, where there are guys on the motorcycle in a gang. So we did the dance routine, and at the end, the leader of the pack on the motorcycle was Christy, and she had to pretend she was on a motorcycle, and it was hilarious. She had her hands up as if she was clutching handlebars, and she just moved as if she was riding a bike, crashing into some chairs and tables as if she was moving the bike through them."

Dawn said *someone* had to volunteer to be the leader, and Christy didn't hesitate and seemed happier to play that role than to be part of the girls dancing and singing.

Christy's father was an advocate for her involvement in political activities. Dawn says one night she went over to Christy Clark's for dinner on the day of a provincial election. Christy's dad was heading over to the nearby elementary school to volunteer, and he invited the girls along to help. They were about twelve years old.

"The ladies who were in charge were having a problem with this because we were so young. I remember her dad defending us because he wanted us to help volunteer. I thought that was so exciting! We were allowed to help scrutineer, and that was just great."

It might surprise people to learn that Christy Clark did not come across as really opinionated or competitive in elementary school. "There was a shyness about Christy for a long time. She didn't stand out." Dawn says that Christy really worked hard to come out of her shell and be able to assert her opinions with passion. "She was so vocal, especially in junior high. That's when she really started practising her conversation skills."

Dawn recalls seeing those skills put to use unexpectedly. "My mom, Marie, and her friend Sue would go out for dinner regularly to a local restaurant, and her friend Sue was very opinionated about current events, so they would often have lively discussions. I remember one night when Christy and I decided to join them. We were talking about an issue, and I could sense that Sue had a strong opinion, and Christy was arguing the opposite point. They broke out into this discussion that started out really amiable and friendly, and I could hear that Sue's voice was starting to rise and get a little loud, and I could tell that Christy was so comfortable in her knowledge of the topic that she didn't need to raise her voice. For Christy, this was really stimulating."

Christy Clark knows that she is often underestimated or misunderstood because of her personality. However, people may be surprised at the importance she attaches to attitude when it comes to achieving goals, something she gets from her mother.

"Because I smile and I have what people would call a sunny disposition, it's easy for my critics to characterize me as less than serious. Imagine in your head a picture of a serious person, this person is never smiling, as if people don't think when they are smiling, or as if optimists don't have a care in the world. I think this idea is short-sighted. I think you can't get things done if you are brooding and pessimistic. Those people are by necessity always looking backward."

She has very strong opinions on the need to maintain a positive outlook. "People who know me, I would hope, will say I'm somewhat smart and I'm optimistic. If you are looking forward and you want to get things done, I think you have to have hope. If you are hopeful, then you know you have a chance to shape the future, you have the chance to make things better."

Clark's childhood environment was a loving, safe place that enabled her to explore all opportunities. She remarks on how different life is today for young people, who are surrounded by technology and outside messaging. Clark remembers her childhood technology: a black-and-white television that was in her parents' bedroom. That was the only location that the television was allowed, and her mother had very strict rules about watching it.

"My siblings and I would fight over which TV shows we would watch after school," Christy says with a laugh. "TVs were pretty new! We weren't allowed in our parents' room without permission. My mother exerted an *iron grip* on everyone's TV time. It was a powerful incentive to stay home sick from school because then you could lie in your parents' bed and you could watch *Sesame Street* and *Electric Company*. Still, I almost never got to stay home from school sick. My mother was a physician's daughter. My grandfather *never* thought anyone was sick, ever, so my mother followed that plan.

"The TV, when we got the colour version, was moved downstairs to what we called 'the TV room.' And that was a big deal. My brothers wanted to watch *Hogan's Heroes* and my sister and I wanted *I Dream of Jeannie*, and we fought about it like cats and dogs, until my mother put a schedule on the wall, and it was followed rigidly. This meant that some days our friends wouldn't want to come over to our house because it wasn't our night for TV."

Christy Clark had, in her father, Jim, a really important role model. She says her dad truly loved his job as a teacher. "He was, in a classic way, called to teaching. My dad was a real connector. People loved to be around him. He was sincere and engaged with people. When he died in 1995, I had letters from former students who saw his obituary in the paper and searched me out to say what a difference he had made in their lives. My dad would have felt that was the greatest tribute of all. He was the teacher who was always coaching the teams, staying after school, getting to know the families."

Did her father's heavy involvement in schooling interfere with his family time? Christy says no, although her mother complained he should have spent more time at home. "I think maybe she was actually complaining about how much he was contributing around the house. That's how I perceive it as an adult: the never-ending debate between husband and wife about house chores."

Jim Clark's role as a guidance counsellor affected many people. One student who benefited from his guidance was Global BC TV reporter Keith Baldrey, whom we will meet in chapter 7. Keith attended Burnaby South Senior Secondary with Christy Clark's oldest brother, John, and although they shared a few classes, they were not friends. However, Jim Clark helped Keith Baldrey with suggestions about going into journalism. Keith has a story that gives a great insight into Jim Clark's personality.

"In the 1975 election, Jim Clark ran as the BC Liberal candidate in Burnaby, and I helped organize an all-candidates debate with Jim

Lorimer, the NDP incumbent, Elwood Veitch for the Socreds, and Jim for the Liberals. Jim was very much 'by the book,' and I remember that he tried to be the mediator between the Socred and NDP candidates." Keith laughs, remembering Jim's comment. "'We just have a philosophical difference here,' he said, and Elwood got really mad. Like a guidance counsellor, Jim was trying to keep peace at a political meeting! The riding of Burnaby–Willingdon was really polarized, more Socred. Veitch won and held the seat for a long time."[4]

Gordon Wilson also remembers Jim Clark well. "He was the first person I met in Burnaby when I was the newly minted leader of the Liberal Party. He said, 'Are you going to stick with it?' And I didn't understand what he meant. Later I realized he meant this wasn't something you do overnight; it was going to be a long haul. He has seen leaders come and go, and he wanted to know I would stick with the job regardless of the challenge."[5]

For all her involvement in politics, Christy Clark had no real political ambition in her early years. "I don't think I ever thought of politics, but in high school I considered being a lawyer. My parents never pushed me to figure everything out, maybe because I was the last of four and they'd given up." She laughs at that thought.

When asked what defines her, she says, "I love engaging with people. There is always something interesting to discover in everyone you meet. What I have noticed about people is that they want to be seen for who they are. If someone comes and asks for help—for example, homeless people—I always talk to them, ask their names. When I worked at CKNW I got to know the regulars, their names, and where they lived. I never really thought about it, until one day some guy came up to me and said, 'You know, lady, I see you every day talking to us, and that's what we really want, is to be seen.'"

Her best times are days with her son, Hamish. "We all know as parents that our time with our kids is short, and I just treasure the time we have together. I am watching him change all the time, and that's exciting, and being able to spend the day together is great."

As busy as life is for a premier who is also a divorced mother, Clark weaves the lessons of her life into her job. Clark's mother's lessons about nature are always with her. Sometimes, the lessons of nature occur in a moment, in a day heavy with governance and politics.

"Outside the back of the legislature there is an old deciduous tree, and the other day when I walked out of my office, through the door that was put in to escape without notice, the sight of this big, amazing tree really struck me. I stood there for a minute and just took the tree in, in all of its glory. I'm sure the staff person with me thought I was crazy, but the beauty stopped me in my tracks."

TOP, LEFT Christina Joan Clark as a baby. COURTESY OF CHRISTY CLARK

TOP, RIGHT Christy in a wheelbarrow with her sister, Jennifer, pushed by their grandfather Johnny Clark. COURTESY OF CHRISTY CLARK

BOTTOM Bruce and Christy Clark on summer break in the Gulf Islands. COURTESY OF CHRISTY CLARK

TOP Christy on some floating wood on the beach at the Gulf Islands retreat.
COURTESY OF CHRISTY CLARK

BOTTOM Christy wearing a fake moustache and smoking a pipe on the patio.
COURTESY OF CHRISTY CLARK

OPPOSITE, TOP The Clark family, circa 1970: Christina, Mavis, John, Jennifer, Jim, and Bruce. COURTESY OF CHRISTY CLARK

OPPOSITE, BOTTOM Christy Clark's parents, Mavis and Jim Clark, on the Gulf Islands with their canoe. COURTESY OF CHRISTY CLARK

OPPOSITE, TOP LEFT The Clark family: Christina, Bruce, John (back row), Mavis (seated), Jennifer (front row), Jim (standing, back row). COURTESY OF CHRISTY CLARK

OPPOSITE, TOP RIGHT Jim Clark as a BC Liberal candidate with the slogan "IT'S TIME TO END ARROGANCE." COURTESY OF CHRISTY CLARK

OPPOSITE, BOTTOM Christy Clark with her mother, Mavis, and sister, Jennifer. COURTESY OF CHRISTY CLARK

ABOVE, LEFT Christy Clark dressed up in costume. COURTESY OF CHRISTY CLARK

ABOVE, RIGHT Christy Clark with a friend on the Gulf Islands, wearing her favourite Ontario Place hat (inside out). COURTESY OF CHRISTY CLARK

TOP Christina and Jennifer enjoying a sunny moment with their grandmother Audrey Bain. COURTESY OF CHRISTY CLARK

BOTTOM, LEFT Hanging out with siblings and friends on summer holidays. COURTESY OF CHRISTY CLARK

BOTTOM, RIGHT Christy Clark in junior high school. COURTESY OF CHRISTY CLARK

You will be loved the day when you will be able to show your
weakness without the person using it to assert his strength.

CESAR PAVESE

CHAPTER 4

A PARTNER
AND POLITICS

ONLY ONE MAN has managed to slow Christy Clark down long
enough to be her partner, and that is Mark Marissen. They were
brought together through their shared passion for politics, and it
was its heavy toll on their schedules that pulled them apart. Still, if anyone is going to have an ex-husband, they should use Marissen as a model.

"I disagree with a number of things she does, but I am not the person
who has to make the decisions. When I agree with her, I let people know,
and when I disagree with her, I keep it to myself... mostly."[1]

Political junkies will know the name James Carville from his role in
the Bill Clinton presidency. Carville is a brilliant political maniac who
doesn't miss a thing that happens in politics or business. Mark Marissen
is, to me, the better-looking, Canadian version of James Carville. Marissen can walk into a room and size up the dynamics like a hunting dog
can smell a fox, and create a political win when others are screaming in
despair. Sometimes he talks in rapid, small sentences, expecting people
to follow along, and laughs at his own insights before moving at light-
ning speed to his next comment.

If Marissen lived in Washington, he would be a wealthy political
operative; in Ottawa, he would do very well financially. Marissen works
in private business in Vancouver, and is a political outsider.

In the 2013 provincial campaign, he was excluded from daily management, and on Christy's insistence, was put on the volunteer executive committee. For the first time, he was not the top dog calling the shots in her campaign. I saw how much he did from outside to support her, and I credit some of her win to the opportunities he helped create, even though few people know about them.

Mark Marissen was born on July 26, 1966, and raised in St. Thomas, a city outside of London, Ontario. He is from a family of Dutch immigrants, and according to one of his best friends, Derek Raymaker, his mum often struggled with finances and other logistics as she raised four boys on her own.

Derek Raymaker studied with Mark Marissen at Carleton University. Raymaker's path took him into journalism and writing. He has seen Marissen's political work for decades.

"For Mark, the Liberal Party is his first love and his passion until the day he dies, and everything else, with the possible exception of his son, will be number two. He is willing to take a bullet for the party."

Raymaker provides this insight into Mark's approach to politics: "Mark loves the institution of the party, and would do anything to protect it. He sees the institution as bigger and more important than the leaders or the elites. He's not scared at all to take on the leadership of the party if he feels he must. A lot of this has affected his reputation within the party. He truly believes that the Liberal Party, and all parties in general, are only as strong as the members that support them. For example, when I was in university with him in the eighties, on issues like Meech Lake, he would speak out against the party based on his beliefs."[2]

Marissen and Raymaker became close friends almost from the first day they met, in 1985.

"It was early in September," says Raymaker, "and we had the same first-year PoliSci class at Carleton. I remember him being very much a Liberal Party activist even back then, wearing the shirts, et cetera. By Christmas, he was more or less running the Liberal club. He was a whirling dervish of activity, loved to sign up new members and organize functions."

Raymaker and Marissen became roommates for a year in 1987, sharing an apartment with another guy on Elgin Street. Raymaker describes this time as "the height of Mark's power in the Carleton Liberals." Of course, being a whirling dervish of political activity does not leave much time for studying.

Marissen was so busy with politics that he was seldom attending any political science classes, and ended up with mediocre marks at school. He knew he had to take drastic action if he wanted to actually finish his degree with the kind of grades he needed for law school, his goal at that time.

That summer, Marissen hung out with a bunch of Young Liberals from western Canada who were in Ottawa for the "Best of the West" program, run by MP Dennis Mills. This inspired him to find out more about BC; he had not been west of Windsor at this point.

Says Marissen, "I decided to go where there were no Liberals. I had never been to BC and went there in the summer to find a university to attend in the fall."

He had missed the deadline to attend UBC and was able to make it into Simon Fraser University (SFU). This missed deadline proved to be fateful, since it put him at SFU at the same time as Christy Clark was attending.

At the time that Marissen began his classes, he was determined to devote all his time to studying. It didn't take too long for that resolution to waver. He remembers reading the school newspaper in the cafeteria, and a story caught his eye about Christy Clark, who had been elected president of the student council and then been disqualified. He found it interesting because at Carleton, he had been disqualified from the board of governors over a technical issue on filing papers. The article said she had left posters up past the deadline, and after the election her opponents filed a grievance.

When Marissen did his homework on the story, he found out that Clark had campaigned on an issue that was part of the motivation for challenging her win: there was a student store on campus, and she thought the jobs at the store should go to students instead of university workers. "She was the enemy of the union for wanting this," says Marissen. "So after she won, the union reps and NDP alleged that she had cheated to win."

It turned out that the electoral standing committee that made decisions on complaints had representation from the union on the committee because the student elections could affect their working conditions. It was the first time Mark had seen the NDP "running a government." His analysis?

"She didn't stand a chance."

The SFU student-politics controversy followed Christy Clark into the 2013 election, when the NDP raised it repeatedly as an example of bad character and cheating.[3] In April 2015, the *Peak*, the SFU student paper, ran a retrospective that showed it was more politics than cheating in that student election. Andrew Tomec wrote the front-page story in April 1989 dealing with Christy Clark's fall from the presidency, and he is quoted in the 2015 story by Cecile Favron. Tomec says Clark represented a slate of candidates called "Unity" made up of moderates and right-wing students, while her opponent, incumbent Paul Mendes, represented the Grassroots slate, the left-wing that evolved out of its radical seventies roots.

Clark's narrow win as president shocked Grassroots students because no Unity candidate had ever won an executive position before; the SFU campus considered itself a "hotbed of revolutionary thought." How did Clark win? Apparently her position of advocating for the students made the difference. Tomec says that is because most SFU students had been apathetic to the work of the student society. The society had spent money on left-wing causes including labour groups and politics in Central America. Clark, in her campaign, suggested this money should be spent on student causes or not at all. As Marissen mentioned, she also focused on student causes like jobs.

In the article by Favron, she quotes Unity member Gerald Christendom from a letter he sent to the student paper in May of 1989. In the letter he said, "It didn't take long... to begin scheming to discredit Christy Clark's election victory." In fact, the student society found that the Unity slate candidates had broken rules, and assessed a fine of sixty dollars. They sent a demand letter to Clark for payment within twenty days, even though the rules allowed for thirty days. She paid twenty-nine days after the fine was assessed, and on that basis, the committee members declared Clark's presidency void.

"I'm a student. I'm not made of money," said Clark at the time. Clark tried to put together some money to fight the decision in court, even saying she would sell her fifty-dollar car; however, with court scheduling and campus politics, the decision held.

When Marissen met Clark, he raised the student politics immediately. He told her he was sympathetic about her disqualification from the presidency and mentioned that he had also been disqualified from the board of governors at Carleton on a technicality. He still remembers her response, and it still makes him laugh. "The lefties got you too!"

He had to tell her that, in his case, it was his own fault over paperwork, and had nothing to do with left-wing ideologues. They didn't have that kind of power at Carleton. Their first meeting started a strong friendship, and they had many long conversations and debates about politics.

Mark was in an eastern political operative's version of *Alice in Wonderland* when he landed in BC, and Christy Clark's approach to politics was a good example of it. He had heard she was a Liberal who was friends with Social Credit people. He was baffled. "But that made *no sense*, how could she be friends with them?"

Not only did he hear that she had made friends with supporters of these right-wing parties, but as he had a chance to spend more time with political students on campus, he found that she had made a coalition with the students who were Socreds and Conservatives.

"I thought this was completely upside down. We are supposed to *fight* them."

Instead, he saw that the Liberals, Socreds, and Conservatives were banding together to fight the NDP.

"This was just as confusing to me. As far as I knew, in all my political involvement, New Democrats were nice people that no one paid any attention to. They had no chance at power. We were not supposed to take them seriously, much less take time to fight them."

Laura Miller, later to become executive director of the BC Liberal Party, is also from eastern Canada. "In BC there is the free-enterprise coalition," says Miller, "which is unique, and although you don't want to paint everything with the same brush, in BC people have this really ingrained 'can-do' entrepreneurial spirit. It's a completely different energy and vibe than Ontario. I don't want to say "wild west", but there's a feeling of opportunity and optimism."[4]

Marissen was drawn into the student scene at SFU, despite the fact that he was supposed to be there to focus on studies and leave politics aside for a year.

"I was very curious about Christy Clark and this strange coalition and the student president controversy because all of it was this weird new world of BC politics." He needed to understand it better, and his natural love of politics drew him in. Marissen says he found Clark funny and passionate, and he was swept up in the new political environment. He talked about his girlfriend to some of his friends at Carleton and told everyone they just had to meet her.

Despite his best intentions, he became heavily involved in politics. Marissen and Clark's relationship strengthened over time, although he had to go home after he completed one year at SFU as a visiting student. Clark stayed in British Columbia, where her involvement with the Liberal Party of BC expanded. Marissen had helped out Jean Chrétien's leadership campaign from British Columbia and moved back to Ottawa after the school year was over to throw himself fully into the action.

The political partnership of Marissen and Clark moved into the national Young Liberals executive, and it was a busy time. Marissen was soon the campaign manager of the Chrétien Youth slate when Jean Chrétien was seeking the leadership of the Liberal Party of Canada. Christy Clark was on Marissen's slate of candidates in the position of external vice-president, based out of BC. The slate that Mark put together had all but one of their candidates elected. Mark was doing this while also working on his political science degree, and after he finished his degree pogram, he became youth director of the Liberal Party of Canada.

Mark and Christy ended up living in two separate places. They talked on the phone every day and saw each other every couple of months, usually when one or the other of them was attending a political meeting or convention. Mark would fly out to BC whenever there was a good seat sale, and they always intended to be together. Christy was hoping and planning to move to Ottawa, but in the fall of 1987 the Liberal Party of BC acclaimed a new leader, and priorities changed.

Under the leadership of Gordon Wilson, there was an increasing amount of work to do in BC, and more dreams to dream. This work and these dreams increasingly captured her focus and efforts for a few years, leading up to the 1991 election and just beyond, and kept her in BC long after originally planned.

Raymaker remembers Marissen and Clark living three time zones apart. "They had a long-distance relationship from 1990 until 1993, when she moved to Ottawa to work with the new [Chrétien] government."

Raymaker became Marissen's roommate again in 1990 after he moved back from SFU. At that time, Marissen was in a serious relationship with Clark, and Raymaker had a chance to get to know her well.

Raymaker remembers the first time he met Christy Clark, in 1990 at the federal Liberal leadership convention in Calgary. At this point, he was a budding journalist covering the story for the *Calgary Herald,* and Marissen was back at Carleton. Raymaker knew that in Marissen's one year attending SFU, he had met up with this mystery girl.

"[He] had reported back to us that he was quite in love with her, she was a very impressive person in quite a few ways. I was looking forward to meeting Christy.

"It was at a Young Liberals party, a social event, at the stampede park. Mark introduced us. I was instantly impressed with her. She's quite electric. She has a bright smile, looks you straight in the eye, with a disarming kind of charm. She can break down your barriers quite quickly with wit, humour, and general good cheer."

Raymaker observes a quality about Christy Clark that is often reported by people who meet her, whether through work or socially.

"When you first meet her, you feel like you've been friends with her forever and known her forever. She wants to know what *you* are all about. What are you up to? What's your family up to? What are your hobbies? She takes a real interest. Before you know it, you are talking more about yourself than about her.

"She has huge energy. She's very memorable. I was happy to have met her. I thought Mark was a very lucky man. I thought he had met his match with Christy."

Life, he concluded, "ultimately means taking the responsibility to find the right answer to its problems and to fulfill the tasks which it constantly sets before the individual." One could make a victory of those experiences, turning life into an inner triumph ... It did not really matter what we expected from life, but rather what life expected from us.

DAVID BROOKS, ON VIKTOR FRANKL

CHAPTER 5

EMERGING ENERGY

ON HALLOWE'EN NIGHT in 1987, Gordon Wilson was acclaimed as leader of the Liberal Party of BC. He was a professor of Resource Economics at Capilano College (later Capilano University) in North Vancouver and was a director on the Sunshine Coast Regional District. He lived outside the Lower Mainland, on the Sunshine Coast, where he had an organic pig farm. Wilson had run as a candidate for the Liberals in the 1986 election, and this motivated him to run for leader. He was thirty-eight years old.

Wilson had a mandate for change, and at the same time that Wilson was changing the direction of the Liberal Party of BC, Christy Clark was changing the direction of her life. Her long-time friend Dawn Clarke noticed that Christy spent a lot of time developing her public speaking skills in small groups of friends. "She went from being fairly reserved, she was a quiet and friendly girl, and as she honed her skills, she was able to speak out more."

Christy Clark spent time in Europe, discovering the world outside Canada, and came home with a new appreciation for her home country.

"I always felt she was very brave to go off to study and work in Europe by herself... Even though she had many wonderful new experiences,

I also got this sense that she was lonely and spent a great deal of time in self-reflection. She came back from that experience with a stronger sense of self and renewed purpose."

Dawn says that, during Christy's Europe and university years, they lost touch a bit, and Dawn noticed Christy's growing interest and passion for working with the Young Liberals. Although Dawn didn't share Christy's political motivations or interests, she was happy to see her friend's personal growth.

"I wanted her to succeed. I knew she would, and she did. Personally for me, it didn't matter who she was, or how successful she became. The only thing that mattered to me was our enduring friendship, as I really valued it."

Dawn felt that Christy's background gave her the confidence to go out into the political world and fight for her beliefs. "She was raised in this wide open, democratic family that encouraged debate and strong opinions ... and early on, she was able to hold conversations and lively debates without getting too emotionally engaged."

Social gatherings revealed Christy's growing interest in topics that her girlfriends didn't always share. "Sometimes on the weekends, a group of us girls would get together, and most of us would be discussing boys, movies, hair and makeup, typical teen stuff, but Christy would be on about politics or current events ... She used to bring the *Globe and Mail* when we would meet for coffee, and my other friends were reading *People* magazine. She was very much a part of the circle, but she was also [developing] her skills and her interests, which were somewhat separate from us. She was never arrogant or pushy, it was just who she is."

Even though many fellow Young Liberals didn't see early signs of leadership, Dawn saw her friend driven by passion. "She knew she had something of value to share, something to say, and she knew people would listen, and so her voice got louder. We would be in a situation, any situation, and if she felt she had to defend herself or stick up for someone else, she would not hesitate ... We would be at a get-together on the weekend, and Christy would be tucked away in some corner or, more often, in the centre of it all, debating some hot issue with someone. I would often feel frustrated, roll my eyes, and think, *Isn't she interested in having a good time?* However, Christy *was* having a good time, the best time actually. She was really lit up during those occasions, and it was evident that she contained the skill, fervour, and knowledge to have a go with anyone."[1]

Another friend has good observations of Clark from their first meeting, in 1991. As a senior civil servant today, Athana Mentzelopoulos is careful in her commentary about the premier. They met when Mentzelopoulos was attending graduate school at the University of Victoria. Even though Mentzelopoulos was never involved in politics, she had been hearing about Christy Clark for years because she attended Carleton University with Mark Marissen and had worked with him on the student newspaper. Mentzelopoulos says that Marissen and others who knew Clark would talk about her quite a bit.

"I had to remind them that we hadn't met. People kept saying that I had to meet her. I was pretty absorbed in post-modernism, a bit of a delicate flower, my graduate degree in philosophy and political science. I was afraid to meet her. People talked about this dynamo."[2]

Mentzelopoulos remembers her first interaction with Clark vividly, because she made an impact when they finally met in Victoria. "A friend was visiting from Ottawa. We all had dinner down on Wharf Street, and my boyfriend and I both walked away talking about how charismatic she was. We were both overwhelmed by her."

It was this energy that Clark contributed to the Liberal Party of BC as the new leader began serious organization of the party. Gordon Wilson was not content with the party being on the sidelines. He was intent on electing MLAs and raising the party's profile. Wilson entered a party burgeoning with feisty Young Liberals, not all keen on an academic pig farmer from outside the city; Christy Clark and Mike McDonald were immediate recruits.

Says McDonald, "This was a very good time to be engaged in politics. The issues were substantial. I remember arguing at length in the club about Meech Lake and the issue of "distinct society," and it sounds like nothing now, but there was a lot of passion about this issue at the time."

In 1987 the Meech Lake Accord was a widely endorsed agreement between Brian Mulroney's Conservative federal government and the provinces, and it was seen as a strong step forward in bringing Quebec into the fold as an endorser of the Canadian Constitution Act of 1982. For the Accord to succeed, it needed the unanimous assent of all ten provinces. Former prime minister Pierre Trudeau became the first vocal critic of the Accord's "distinct society" clause and declared the document a "capitulation to provincialism," and the debate escalated.

"The Young Liberals felt engaged," says McDonald. "It was more real than what we were studying in political science. It was a very divisive time for Liberals. Turner supported Meech Lake; Chrétien did not. Our new leader, Gordon Wilson, took a lot of flack from the federal Liberals for his opposition. I think Young Liberals back then were pretty much all anti-Meech; there might have been a couple of supporters, but most of us were pretty upset about it."

Christy Clark was one of the supporters of Meech, as was I. My position flowed from our federal leader John Turner's position, but my position on Meech did not deter me from working hard for the new leader, Gordon Wilson, on building the provincial party. It did not appear to deter Christy Clark, either.

In the end, Meech Lake was defeated in 1990 by Elijah Harper, the MLA for the riding of Rupertsland, Manitoba, a member of the Red Sucker Lake community, and the first Treaty Native person to be elected to office, in 1981. Harper stood up in the legislature holding an eagle feather and initiating a speech that ended the prospect of unanimous consent for a vote on Meech, which killed the deal.[3]

Until its defeat, Meech Lake was an issue that separated the loyalties of the Liberals who supported Meech Lake federally and provincially and had used Wilson's opposition as a reason not to work on the provincial party. What this meant, in a practical way, was that the troops on the ground between the 1986 and 1991 elections remained thin, even with huge efforts by the new leader and his growing team.

This meant that Vancouver-based workers like Mike and Christy were very well positioned to help with the 1991 election.

Suffering, oddly, also teaches gratitude ... People in this circumstance also have a sense that they are swept up in some larger providence. Abraham Lincoln suffered through depression through his life and ... emerged with the sense that Providence had taken control of his life, that he was a small instrument in a transcendent task.

DAVID BROOKS

CHAPTER 6

THE 1991 ELECTION

T HE FOCUS ON the 1991 election and its unexpected breakthrough for the BC Liberal Party is often on Liberal Leader Gordon Wilson and his debate performance, which triggered an unstoppable momentum that elected seventeen new MLAs. Wilson will tell you that, in order to have a chance for this breakthrough moment, it took a lot of backbreaking work from a dedicated team. He will also tell you that Christy Clark was a key player on that team.

Wilson first met Christy Clark in 1987 after he became leader of the party. "The first memory I have is at a Young Liberal event at the SFU campus. I came in as the new leader to meet members and speak. My first impression of her was that she was, in the context of high school days, the kind of girl most guys would want to hang around with, but she was out of their league. She was vivacious, bubbly, talkative, and seemed very outgoing. She was a bit tomboyish and a very attractive woman. She exuded a tremendous amount of self-confidence. The thing that struck me most of all was her smile. She was always smiling."[1]

When they met, they talked politics, and specifically, they talked Liberal policy. "I was interested because I knew her father from past political campaigns. I had come into the Liberal Party as a neophyte. My sister Susan [French] was more engaged, and it was through her that I had the introduction to the Clark family when I was leader. We were trying to build a base. I was interested to see where Christy fit into the federal/provincial Liberal split. So many Liberals were Socreds provincially. She seemed to be Liberal all the way through, a very centrist ideology. Her ideas of fiscal reform stemmed from her father, and his 'don't spend what you don't have' philosophy."

After Wilson became leader, he initiated a round of policy discussions in order to develop a comprehensive platform to take to voters in the next election. It was also a way to engage new members and to make people feel relevant to the process. He ended up working with Christy Clark on quite a few policy discussions.

"We developed policy on everything from education to social services to the more innovative policies like fixed election days and making parliament more relevant. She seemed to be one of a number of people who were quite interested in drilling down to the details on those issues. At the same time, she seemed to be having a lot of fun doing that. I don't think she took life too seriously. She was one of the YLs you knew was out there not only working hard but also having a lot of fun at the same time."

Mike McDonald was an active volunteer when Gordon Wilson became leader and served as BC Young Liberals president from 1988 until 1990, when membership was growing and issues were dynamic. He refers to Christy as a trusted confidante during this time, even while she was engaged in student politics at Simon Fraser University.

"At one point, we had tried to raise money for the broke Liberals. I recall running a volunteer phone bank where we had some success cold-calling members for funds. Christy, of course, was a natural and made sales like no other."

In addition to Wilson asking every party member to help with fundraising, he was determined to split the provincial and federal parties into two separate entities. In part, he did this because, every time his team raised money, he had to fight to get it from the Ottawa office. The Young Liberals were a big part of the team who worked hard to help split the party, and this was successful in time to make the BC party its own entity before the 1991 election.

It might seem counterintuitive to say that the Liberal Party had to be split in two to have a chance of doing well in the election, but for a breakthrough, there had to be a united focus. With the party holding members who were both federal Liberals and provincial Socreds, you could not even hold a safe strategy meeting.

At the May 6, 1989, Liberal convention in Penticton, a huge crowd turned out to fight over whether or not the Liberal Party in BC should be split into federal and provincial wings. It was a bitter fight, with many lifetime Liberals passionately opposed to any division. This was the day that Floyd Sully, businessman and computer expert, took over as president of the Liberal Party of BC. A number of party members from Richmond ran for the executive, and Floyd was elected with a few others.

"We wanted to take control of the executive," recalled Sully. "We ran with the slogan *Divided We Conquer*. We were big on splitting the party, and even then it took two years to do it."[2]

Gordon Wilson provides this perspective: "The reason we needed to split the party is because the federal party controlled the money. They were not prepared to allow sufficient funds into the provincial wing to provide resources for a credible campaign.

"It was incredibly difficult to split the party because most of the people who were delegates to the convention were federal Liberals and provincial Socreds. We wouldn't quit our efforts, though, and in the end, I think we succeeded because we were so damned obnoxious that they wanted to get rid of us. They thought we wouldn't amount to a hill of beans anyway.

"The people who cheered about basically being kicked out included a lot of Young Liberals, and keep in mind, a lot of them were really frustrated because they had worked very hard to raise money only to watch it disappear into Ottawa and we [the provincial party] would still be broke."[3]

Many of the small army of Young Liberals who were engaged in politics in the late eighties and early nineties in British Columbia are still involved in public service today. In fact, BC cabinet minister Todd Stone was a Young Liberal volunteer in the 1991 election, and after years in the private sector, returned to politics to run for the Liberal Party under Christy Clark in 2013. Christy Clark and Mike McDonald were responsible for helping with candidate recruitment. That is a lot of responsibility for two people who were in the group often referred to as "the kids." They took it on happily and began preparing for the general election.

In mid-1991, BC politics was in a state of considerable turmoil. The Social Credit Bennett dynasty ended when Premier Bill Bennett left politics to return to private business in 1986. Bill Vander Zalm won the leadership of the Social Credit Party and the October 1986 election for the Social Credit Party, and then Vander Zalm's term had a series of controversies.

Several scandals plagued the government, and unrelenting headlines and news stories, concerning either cabinet ministers or the premier himself, were seen by voters frequently. After the controversy concerning the sale of the Vander Zalm family property known as Fantasy Gardens, Vander Zalm resigned in April 1991.[4]

After Vander Zalm's resignation, the Social Credit Party held a leadership race, contested by Mel Couvelier, Grace McCarthy, and Rita Johnston. All three had served as high-profile cabinet ministers, and many expected Grace McCarthy to win. When Mel Couvelier dropped out to support Rita Johnston, she won and became BC's first female premier. She inherited a party with considerable challenges, and her government experienced controversies, leaving the impression that some key cabinet ministers were unethical.

The difficulties in the Social Credit Party created an opportunity for a change in government. The NDP were led, at that time, by Mike Harcourt, the former mayor of Vancouver. Nicknamed "moderate Mike," he was not as scary to the BC business community as some of the more socialist NDP members, and he was doing well at articulating popular policies such as stronger protection for the forests, cleaning up pulp mill emissions, and strengthening worker protection. Harcourt continually reassured the business community that he would be able to govern without any surprises.[5]

In the Liberal Party, we all sensed an opportunity, because many of us believed that voters would look for an alternative to the NDP *and* the Social Credit Party. This required a huge amount of volunteer work because the party had no money. In fact, many of us went beyond volunteering and actually covered every cost for travelling or phoning or recruiting supporters. Some volunteers were more motivated than others, and some were more effective than others.

The election of Floyd Sully as president of the Liberal Party of BC in 1989 was likely as important to the administration of the party and its preparation for the 1991 election as Wilson's was to leadership and political momentum. Sully was a man with strong business credentials, and his ambition for the party was as big as Wilson's. Because he owned his own business, he could be flexible in his allocation of time and investment of resources. He believed that Gordon Wilson would make an election breakthrough, and he worked tirelessly at the centre of the administrative team to make sure the key elements were in place to support it.

This was a good basis on which to build the new party, using Wilson's leadership, Sully's experience, and the Young Liberals' endless energy.

According to Gordon Wilson, Christy Clark "was instrumental in recruiting candidates. She and Mike McDonald and Todd Stone were part of the group of Young Liberals who were actively out helping to build the party under my political leadership and the party leadership of Floyd Sully. I usually talked to the Young Liberals in a group, and not all of them were that supportive of me as leader. Quite a few thought that it was all an exercise in building for the next leader. Many of them thought I would not be able to lead the party to any elected seats. I don't know what her thoughts were, except I noticed that she worked very hard as part of the team building the party."

Gordon Wilson had tasked Sully and the executive with making sure there was a full slate of candidates in the election. This was an enormous job because there were many parts of the province where the Liberal Party had almost no members. This was before the Internet, before cell phones, and before Facebook or instant messaging. British Columbia, a province that is the size of most of Europe, was a huge territory to cover, and the only way to do it was to get on the road. The leader had travelled around the province nonstop for almost four years. Now was the time to "close the deal" and get candidates in place. There was an immediate need for seventy-five candidates in order to be a party that was taken seriously by the voters.

I remember this time period well because I was living in the Okanagan again, having babies every two years, and doing whatever I could to help organize the ridings around me. In 1988, Wilson talked me into running in a by-election in the south Okanagan, and many Young Liberals would do regular road trips to help my campaign. My campaign manager, Graeme Keirstead, was a Young Liberal.

As the 1991 election approached, the leader had asked me to run as a candidate and help recruit candidates for the other Okanagan ridings. I was pregnant with my third baby by then and not that connected to the central campaign, but I heard the buzz every time a new candidate signed on. Floyd Sully remembers the election campaign in full colour, and has some great stories to tell about how Christy Clark and Mike McDonald helped corral the candidates for election.

"Before the seventeen were elected," says Sully, "Gordon gave us our marching orders: we had to get candidates in every riding if possible. The problem was finding candidates in the remote ridings. Mike was president of the Young Liberals; Christy was vice-president. We couldn't have had two people more versed in the policies and better known to the older members. That was the role of the youth, to prod the older members to get involved. They didn't have much in the way of money, and they went out to the Kootenays, to the north, around the Interior; the party was moribund in many of the ridings, so they had their work cut out for them." Sully laughed. "One of the lines they used was, 'If we don't find a local candidate, then the party president will have to run [as an out-of-town candidate], and that's going to look pretty embarrassing for you local members.'"[6]

Mike describes the trip as "a bit of a legend." "We took Clive [Tanner]'s Chrysler van and drove to Prince George, then we worked our way down the province finding candidates. We hot-tubbed with a candidate in Prince George; we dragged a mustard factory owner onto the ballot in Quesnel; we found one of the few French guys in Williams Lake and he ran; a college friend in Kamloops was on the ballot, and we found a running mate in Clearwater for him. We had soup with iconic Salmon Arm Liberal Basil Studer, who rustled up the town surgeon. As we went, we shook the tree."

McDonald makes it pretty clear how the team work was divided up: "I was the one who found the targets and lined them up. Christy was the closer."

They had seventy out of the seventy-five ridings, and the clock was ticking on the deadline for nominations. Party President Floyd Sully realized he would have to run in the riding of North Vancouver–Lonsdale.

"The papers had to be filed by eight that evening. We didn't have enough members in the riding to get the signatures. Once again, the Young Liberals came through. They said, 'We will go door to door to get the fifty signatures.' Everyone said they were crazy and it couldn't be done, but those young people went out and they knocked on doors and explained what we needed, and were surprised at the response. People were very pleased that they were being asked to participate in the political process, and some of them signed even if they weren't Liberals.

"The Liberal Party could not have claimed a position on the CBC [televised leaders'] debate if we hadn't had a full slate of candidates," says Sully. "In my mind, the CBC debate was absolutely critical to our victory. After that debate, every day, the Liberal Party was rising in the polls. It was clear that the debate gave us the momentum forward. We had no other resources. The entire central campaign ran on less than $100,000."

Media coverage of the Liberals' attempt to be included in the debate was carried as far away as Ottawa. Interestingly, speculation was that

Wilson's inclusion on the debate would hurt the NDP and help the Social Credit Party. The *Province* reported:

> Question: Can the same reasoning process lead to a pair of diametrically opposite conclusions? Answer: In politics, yes. Before CBC-TV capitulated last night, Premier Rita Johnston argued that Liberal leader Gordon Wilson should join the televised leaders' debate, while NDP leader Mike Harcourt argued he shouldn't.

> The reason: Strategists for both the Socreds and NDP believe that Wilson's Grits will get a good chunk of votes from the group fondly known as "soft NDPers." According to backroom luminaries for both parties, the more exposure Wilson gets, the better for the Socreds, the worse for the New Democrats.[7]

Sully and the Liberal team didn't care about which party might be affected; their focus was on having a place on the debate to build their own base.

"All we had was Gordon's performance in the debate," says Sully. "The full slate that qualified us to attend was due to Mike McDonald and Christy Clark going out and doing the hard work. They get full marks for it. We were convinced that if the public could see Gordon with the other leaders, then we would have a chance. And they did, and we were right."[8]

Mike remembers how difficult it was to find candidates and adds that "in today's age of social media, a lot of those candidates would not have held up too well to the scrutiny. But we got our piece done, and Gordon Wilson took care of the rest."

The media were not particularly happy about the last-minute addition of the Liberals to the debate, and press gallery reporter Brian Kieran summed up the general grumbling of the media in a story in the *Province* dated October 8, 1991. He talks about the "meticulously crafted leaders' tour" that had been under way for almost three weeks and the challenges facing the NDP and the Socreds, then states his grievance:

Unfortunately, the debate tonight will be quite different from the one originally planned.

In fact, it won't really be a debate at all. It will more closely resemble an extended, rapid-fire press conference.

We have Liberal leader Gordon Wilson to thank for this.

Wilson has devoted most of his campaign to a harangue against big media for paying too little attention to his low-budget bid for credibility.

This has resulted in his one campaign triumph, insinuating himself into tonight's battle of the titans.

And it sent CBC producers scurrying back to the drawing board.

The format for the debate—as laid out on paper yesterday—is a timekeeper's nightmare.

The only voters who will remember anything will be those who can afford to hire a stenographer.

Apparently, the CBC is attempting some last-minute monkey-wrenching, when it's clear major surgery is needed to rescue this terrible format from being a prime-time waste of time.[9]

In the headlines, the speculation was that the Social Credit Party was on its way out and would have a hard time recovering, while the NDP was on its way in. The negative coverage of the Social Credit's chances was hard on Premier Rita Johnston, who became leader after considerable damage to the party's reputation. Rita Johnston was BC's first female premier, and a respected cabinet minister before becoming leader, but the Social Credit Party had been in government since 1975, and many felt it was time for change.

Vancouver Sun press gallery reporter Vaughn Palmer wrote that, just before the debate, Premier Johnston talked about some of the realities of the campaign:

For the first two weeks of a discouraging election campaign, Premier Rita Johnston remained the gamest contender on the Social Credit team. "I'm told by those working on my campaign that this campaign's going to be a lot of fun," she quipped amid the disasters. "Well, I'm waiting for the fun to start."

"We can say all we want about 'things are great,' but we know we're down, and it's demoralizing for our troops," she told Justine Hunter of the *Sun* and Les Leyne of the Victoria *Times Colonist*.

Particularly successful has been the attack on Socred ethics. "No question, ethics is right up there," Mrs. Johnston said. "I spend far more time responding to those kinds of questions than I hoped I would have to. It is actually more of an issue than policies. The NDP campaign, although I think less than honest, has been very successful in making that an issue."[10]

Vaughn Palmer predicted that what was needed to decide the election was a one-liner, and quoted Premier Rita Johnston as saying, "We're all looking for that 30-second clip." Palmer, in analyzing many key debates and their impact on elections, demonstrated a number of examples of significant changes in election outcomes based on a single moment or line in a debate.

Those who analyze and report debates between political leaders will usually search for "the clip"—a line or brief exchange between the candidates, which sums up who won in a single, televisable segment of no more than 30 seconds...

It sounds absurd, I know, that a debate, with all its hours of preparation and wrestling with the issues, will be summarized, graded and remembered on the basis of a single answer, perhaps even part of an answer. Yet few rules of thumb are as reliable in politics as this one.[11]

Following the debate, there were a number of reports about the performance of Premier Johnston or Mike Harcourt, with a general consensus that Premier Johnston had won the debate, which could boost the Social Credit Party's polling numbers. There was little mention of Gordon Wilson or the Liberals.[12]

Less than a week before the general election, on October 12, the *Vancouver Sun* and Angus Reid released a poll, similar to the one broadcast on the eve of the election on BCTV (now Global BC TV) news, showing a surge in support for the Liberals following Wilson's debate performance.[13] The polls claimed Liberal support was 35 percent to the NDP's 33, while the Social Credit trailed at 19 percent. It was a shocking change.[14]

On October 17, 1991, the Liberal Party of BC elected seventeen MLAs, the NDP formed government with fifty-one MLAs, and the Social Credit Party elected seven MLAs. The Liberal Party of BC became the Official Opposition. Keith Baldrey wrote a comprehensive overview of the election campaign and its results for the *Vancouver Sun*, excerpted here:

> It was supposed to be such an easy election campaign to figure out: The NDP was going to win and the Socreds were going to lose. The Liberals? Hey, who cared about the Liberals?
>
> Well, as it turned out, enough people cared about the Liberals—or at least cared less about the NDP and the Socreds—to turn this campaign upside-down at the halfway point.
>
> Harcourt was confident and relaxed. Johnston was harried and irritable. And Wilson was . . . well, we don't really know because the media really wasn't covering him very much.
>
> By Oct. 8, the NDP appeared to be headed towards a comfortable majority. Nothing stood in their way except a self-destructing, disorganized and demoralized bunch of Socreds.
>
> But that evening, something extraordinary happened. Wilson, like the little brother finally being allowed to tag along for some fun, was able to meet his two would-be opponents in a face-to-face confrontation on live television.
>
> It wasn't pretty.
>
> While Johnston screamed and sniped at Harcourt, and while Harcourt bobbed and weaved and ducked questions, Wilson aimed his sights at both and fired at will. He didn't miss too often.
>
> It was still a two-party race, but the Liberals were now in the game with the NDP—not the Socreds.

Daily polls showed the Liberals gained 20 points almost overnight. Surely, the pundits said, it was a blip and that support would soon drop.

In the last week, the NDP coasted along, still confident but not quite so cocky. The Liberals were giddy with excitement. The Socreds glumly began seeing reality, admitting it was going to be pretty bad for them.[15]

Baldrey's admission that the media hadn't been paying attention was pretty accurate. Following the 1991 election, the media repeatedly placed credit for the Liberal breakthrough not only on Wilson, but on one line he uttered during the debate.

Within the party, however, we all knew better. We understood that it had taken a small army of volunteers focused on building the party in order to take advantage of the opportunities presented during the election, and that it was this hard work and this solid base that elected Liberal MLAs, most of whom had never been elected previously.

It was an exciting time, and a thrilling challenge, and the team was ready to work hard to prove that the voters' trust was well placed. Christy Clark was one of the people ready for the next step.

I will be conquered. I will not capitulate ... he whose life has passed without contest, and who can boast neither success nor merit can survey himself only as a useless filler of existence.

SAMUEL JOHNSON

CHAPTER 7

THE BASEMENT YEARS

MIKE MCDONALD SAYS that after he and Christy Clark finished helping the central campaign with the recruitment of the seventy-one candidates, they were sent to the Sunshine Coast to join the campaign to elect Gordon Wilson as MLA.

"I remember the feeling of unstoppable momentum after the debate. Sign requests, volunteers, money. Christy and I stayed in Sechelt during that time with our good friend Kathryn Choquer, an active Young Liberal. On election night, we watched the results in Sechelt, and it was tremendously exciting."

When the Liberal Party of BC had its breakthrough and elected seventeen rookie members, it was after backbreaking work by a team of people who slogged away for years, and it was a very exciting night for those workers. On election night, the Sechelt volunteers took a boat ride down the coast to join the real victory party.

"I remember that boat ride," says McDonald, "a late October evening that was as calm as you could imagine, and a bright moon. It was a surreal experience to take this quiet, idyllic ride, knowing at the other end it would be mayhem, and our young political lives would be changed."

In 1991, I was one of the seventeen MLAs elected under Gordon Wilson's leadership. I was twenty-six years old, and I was the only MLA from the Interior of the province, elected out of Kelowna, which at that time was still the Social Credit heartland. Because I was pregnant, had two little children, and was the only Liberal MLA with a distance to travel, I did not attend as many organizational meetings as I might have otherwise, and I skipped a few of the get-togethers. It was when we were preparing for the 1992 legislative session, the first one as elected members, that I saw Christy Clark again.

She had been hired as a researcher with the Liberal caucus, and the offices for the Liberal caucus were in the basement of the legislature. The Liberal caucus offices where I was posted were on the second floor of the buildings, and it was a bit of a maze to try to navigate from the MLA offices down to the research offices. I tried it a few times, but often just didn't have time to run up and down because I was either very pregnant, or dealing with a newborn and my first legislative debate session. It was also really awkward for me because many of the Young Liberals who had been hired as researchers were young single people who were my age, and I was an MLA with three small children, so I always felt out of place. As a result, I didn't work closely with Liberal research, and tended to do my own research.

Christy Clark had been hired by Gordon Wilson to work with Richmond MLA Linda Reid as a researcher.

"I first met Christy," says Reid, "in the late eighties when she and Clive Tanner and Mike McDonald were travelling the province recruiting candidates. I was nominated, and the caucus chair, so I would receive regular updates on their progress." Christy Clark provided Linda Reid with research on issues like health, education, social services, and justice. Says Reid, "She was articulate, strategic, humorous, and focused."[1]

Christy looked different from when I had last seen her, but then it had been a few years and a lot had changed. My memory of her while she was a researcher is that she was quiet, worked tirelessly, was intensely focused on her job, and was well respected. Also, her hair was short and darker, and she seemed to be dressing so she would not be noticed. I remember her as being somewhat subdued, a quieter version of the person I had known at the Young Liberal socials. This made sense, of course, since a legislative office is quite a different environment from a political get-together full of young and opinionated volunteers.

Liberal research was loaded with young people who had worked so hard to help with the breakthrough, and of course, they were hired with all of their previous political opinions intact. This was the good news *and* the bad news because, at that time, the Liberal Party of BC had only recently been separated from the Liberal Party of Canada.

Within a year of the election, a national issue created a huge fight within the Liberal caucus, and this fight created tension and divisions throughout the party, testing the Liberals. There were some divided loyalties when BC's new leader of the Official Opposition took a strong position on the Charlottetown Accord, a proposal to amend the Constitution of Canada, with divisions expressed both upstairs (between the elected members) and downstairs (with Liberal researchers). Christy Clark was not one of the problematic researchers. Upstairs, MLA David Mitchell resigned from his position as House Leader, and MLA Art Cowie challenged Wilson's leadership. Behind the scenes, staff were arguing in favour of the federal Liberal position and were very angry that Wilson was one of the leading voices arguing against the Charlottetown Accord.[2]

Gordon Wilson recalls, "I was offside with a lot of the Liberal researchers. The party was split on whether we should support Charlottetown. The federal Liberals had taken a position, and it ran counter to the position I had taken. I was working fairly closely with Pierre Trudeau, Clyde Wells, and Sharon Carstairs. From coast to coast, there were Liberal leaders ringing the alarm bells that this was not the way to build a united Canada. I was heavily lobbied by people in the Liberal research staff, and among caucus, to change my position, but I don't remember Christy participating in any of the lobbying."[3]

It is telling that, for someone with Clark's passion, she was prepared to accept that she worked for the BC Liberal leader as opposed to being swayed by the pressure from Ottawa to follow the federal leader, and speaks to Clark's strong BC roots and perspective. I remember Gordon

Wilson saying, "They keep yelling at me, 'The leader is saying we have to support Charlottetown!' and I finally said to them, 'In this building, I'm the f'ing leader!'"

Senior press gallery reporter Keith Baldrey remembers Christy Clark's time as a researcher because that is when he first spent time with her as a political operative.

"When I first met her, I have to say, she was a bit of a party girl and wasn't keeping herself in the best shape. She smoked a lot, partied hard, and worked hard. It can be easy for young political staff at the legislature to live and breathe the place and to exist in a rather insular world, and I think she probably got caught up in that."

Like so many others who knew Clark in the early nineties, there was little indication of leadership ambition.

Says Baldrey, "She wasn't seen as a player or a potential future politician."

Vancouver Sun veteran political columnist Vaughn Palmer remembers that Christy Clark was a staff person in the BC Liberal Opposition but says there was little else to remember of her from that time and no indication that she would move from a staff position to premier of BC within twenty years.

He makes an interesting note about a possible trend concerning BC political staff. "Another staffer from the nineties is now premier of Alberta. If Horgan ever makes it, that would be three. Should probably pay more attention to staff."[4]

Baldrey believes that the 1991 breakthrough set the stage for Clark's leadership ambitions because it built on her father's political ideas. "If you have a parent who runs for political office, you are naturally drawn into that world. Certainly her activism itself was influenced . . . the party had been mortally wounded by the defection of (former Liberal MLAs) Pat McGeer and Garde Gardom, the Socreds cut their legs out from under them. Her father remained loyal to the party, and she remained loyal to the party, and the Socreds were no longer a factor after Gordon's Wilson's performance."

Baldrey is referring to the fact that, even though a few Social Credit MLAs were elected in 1991, they had disappeared as a political force by the time the 1996 election occurred.

Says Baldrey, "If not for the performance of the Liberals in 1991, Christy Clark in all likelihood would never have become MLA and would not be premier today.

"The 1991 election, like the 2013 election, changed so many things."[5]

The genius of this conception is that as people become more dependent on God, their capacity for ambition and action increases. Dependency doesn't breed passivity; it breeds energy and accomplishment.

DAVID BROOKS

CHAPTER 8

THE OTTAWA YEARS

SOON AFTER THE 1990 federal Liberal leadership won by Jean Chrétien, and following the 1991 BC election, Mark Marissen had a full-time job as youth director of the Liberal Party of Canada. Jean Chrétien was the leader, and there was optimism in the federal party that Chrétien and his team would be able to win power in the upcoming election.

Marissen had a busy summer in 1992, trying to organize young people across Canada into the federal Liberal Party, while the Liberals in BC were arguing about the labour code and the constitution. It is little wonder that Conservatives disliked Marissen; in addition to being an effective organizer for the party, he was quick to take jabs at them in the materials created to recruit youth to the Liberals. He seemed to know how to get under their skin. In a story published in the *Ottawa Citizen* and the *Montreal Gazette*, reporter Jane Taber wrote:

> The Liberal youth are launching a campus recruitment campaign this week. They're sending out posters depicting the "Tory face of Canada" and some other fairly amusing literature to their campus clubs.
>
> It's an intensely partisan set of documents geared not only to increasing the youth contingent at university and college campuses

(according to young Liberals there are about 55 active campus associations with between 50 and 500 members each) but also to getting the youth ready for an election.

The campaign manual, prepared partly by young Liberals' national director Mark Marissen and president Michel Chartrand, is full of helpful hints about everything from "building your team" to packing a room to dealing with the media, especially the inexperienced campus media...

While Marissen and Chartrand are excited about the manual they're even more enthusiastic about the recruitment poster that shows "the Tory face of Canada"—employment lineups, environmental pollution and the prime minister rolling the dice all cleverly put together to form Mulroney's profile.[1]

About a year before the 1993 federal election, Mark was offered a job working for Gordon Ashworth, the national campaign director for the Liberal Party. In order to spend more time together, Christy applied and was hired to replace Mark as youth director for the Liberal Party of Canada, leaving BC and her job as researcher in late 1992. She moved from BC politics into the dynamic hub of federal Liberal politics in Ottawa just as the Chrétien machine started to gain traction and prepare for government.

"We went from living apart for two years," says Mark, "to living together and working in the same office through the 1993 election, with a great outcome in terms of Chrétien being elected prime minister."

Still in their twenties, they were in the midst of a surge of energy and excitement. They were contributing to the success of a party they both supported as true believers. "It was fun," says Marissen with a big grin.

Christy was hired by Liberal Member of Parliament Doug Young, an MP from New Brunswick, and Chrétien's minister of transport. Doug Young was not too popular on the BC coast since he was the "man who closed down the lighthouses," and Marissen recalls that being a contentious issue.

Meanwhile, Mark landed a job with newly re-elected Member of Parliament David Anderson. Anderson was elected as MP in the early seventies for one term, then MLA for one term in BC while BC Liberal leader, and had returned to politics for the 1993 election, serving in cabinet for Chrétien as Minister of National Revenue.

Marissen and Clark were both kept very busy with their work for the new government. Finally, they were together and living in Ottawa, but ironicially, Mark ended up in BC more often than Christy because, in addition to Anderson's work as Minister for Revenue, he was also the senior minister for BC, and there was a lot of work for Mark to do on communications and organization that required travel to BC from Ottawa.

Clark was busy networking in the capital, and Don Guy, who would later play a key role in the 2013 election, first spent time with her when she worked for the Liberal government in Ottawa. Guy says, "Mark Marissen and I were at Carleton together, and we had many mutual friends in the young Liberals and on staff in Ottawa."[2]

Clark made a strong impression. "She was a force of nature. She was the centre of everything. Whatever room she was in, she was at the centre of it in terms of her energy. People wanted to be around her, and she had natural leadership qualities that really shone through. We encountered each other occasionally but weren't close, so I don't know if she had leadership ambitions then."

Long before Christy Clark considered running for leader, Don Guy had noticed her leadership qualities and ethics. "It definitely caught my attention how hard she worked and her ability to get people enthusiastic about participating. This is when she really first struck me as a potential party leader. I told her as much over coffee in 2002 and said, if she ever needed help, I would be there."

Athana Mentzelopoulos encountered Clark while she was working for Doug Young, and she has similar memories of Clark's personality and the effect she had on others. Mentzelopoulos had worked hard to land a position as a civil servant with the new government. She was not political, although she had friends from her university years who were, and, through those connections, met the new staff person from BC.

"I wanted to go to medical school and thought it would look good on my resumé if I had worked for the federal minister of health. I sent [Dave Dingwall] a registered letter every day for thirty days, begging for an interview and saying he should hire me. He did hire me but didn't end up as Minister of Health!... Christy was working for Doug Young. Our circle of friends was the same, and we ended up hanging out quite a bit, and after a year or so we became very, very close friends, and not long after, she left Ottawa."

Mentzelopoulos was able to observe Clark in the same kind of situation that Clark's childhood friend Dawn Clarke did, and it made a strong impact on the young civil servant. "I remember being in awe of her even after we became good friends. She was someone who could look at any political problem and take it apart. I got my job in a minister's office, and I was scared every day. Meanwhile she was calm and, basically, she would hold court at Maxwell's and would engage anyone in debate. She was a smoker at the time, and we could smoke in restaurants, and there was one guy who would get really mad that she was smoking, and she basically kicked him out of the restaurant once."

Clark's love of people and her interest in social settings played well in the dynamics of Ottawa. "She was just fearless. People wanted to be around her. You could never go to her house without twenty people hanging out. Someone would show up with food, and she would just roll with it. People loved her. They wanted to be near her. She was magnetic."

The federal Liberals won government in October of 1993. While Clark and Marissen were living in Ottawa, BC politics changed fundamentally, again, especially in the Liberal Party. Gordon Wilson ended up in a high-profile challenge to his leadership by his caucus members, who challenged his judgement and his honesty about his personal life. Wilson tried, unsuccessfully, to portray this as a bid to remove him as leader by Gordon Campbell, then mayor of Vancouver, and those who had previously run the province under Social Credit.

I was involved in this controversy; there are many layers to this story, and I wrote my perspective in *Political Affairs* (1994). The Liberal Party elected Gordon Campbell as leader in the fall of 1993. Gordon Wilson and

I left the Liberal Party and started a political party called the Progressive Democratic Alliance and became independent MLAs in the legislature. All the other Liberal MLAs remained in the Liberal caucus elected under Wilson's leadership in 1991.

Christy Clark missed this activity because she was in Ottawa and immersed in the new Liberal government. Campbell became leader of the Liberal Party during Mike Harcourt's term as premier.

Campbell began to rebuild the so-called free-enterprise coalition and renamed the party the BC Liberal Party (instead of the Liberal Party of BC), changing its main party colours to Conservative blue from Liberal red.

Gordon Campbell was not an elected MLA when he won the leadership, so Art Cowie resigned his seat, and Campbell easily won the by-election. Shortly after becoming leader, Gordon Campbell travelled to Ottawa for a reception for Liberal staff members, and to make connections with the new government. Campbell was actively recruiting people to run for him, and he asked Christy Clark to help pull an event together. Clark in turn asked Marissen to do it because he knew a lot more people and worked for David Anderson, who was keen to help Campbell. Ironically, Marissen said that Campbell's staff did not want Anderson to attend, but he went anyway.

The dinner was a big success, and according to Marissen, "After the dinner, Campbell asked Christy to run." Clark, intrigued and interested in returning home to re-engage with BC politics, started to consider a run in the next election, and Marissen thought she should explore the idea.

Mark remembers that on their next trip to BC for a family visit, they asked Mike McDonald to do an analysis of the seats available to Christy if she decided to run for the BC Liberals. McDonald focused on Burnaby Mountain.

"Christy and Mark had spent New Year's with me," says McDonald. "And I remember looking at the options. She was from Burnaby, and there were ridings available near where she grew up. I supported her choice of

Port Moody–Burnaby Mountain because it included some Burnaby but was also the type of riding that suited her well—lots of young people, up for grabs (not too NDP), and she could go in and build an organization."

Clark decided to run and, with Marissen's support, moved back to BC in 1995. At that time, she stayed mainly at her dad's home, while Mark stayed at work in Ottawa. Jim Clark, a seasoned campaigner, was a good sounding board as she prepared to follow in his footsteps as a candidate for the Liberal Party in BC.

The truly great leader overcomes all difficulties, and campaigns and battles are nothing but a long series of difficulties to be overcome.
GEORGE CATLETT MARSHALL

CHAPTER 9

RETURNING TO THE COAST

I F A BIG brother is supposed to watch out for his little sister, Bruce Clark is a good big brother. He watches every trend in politics and uses his vast business network to keep an eye on current events and issues, always trying to find ways to protect his sister without anyone even knowing that he is working behind the scenes.

When Christy Clark decided to run for MLA in Port Moody–Burnaby Mountain, Bruce Clark was involved. There are even some indications that he set the entire process in motion while she was in Ottawa, ensuring that she caught the attention of the new BC Liberal leader, Gordon Campbell.

Bruce used his own political network to help Christy Clark establish herself in her first run for office. "I ran for office a couple of times, municipally. After realizing that I wasn't that well suited for it, I had a huge network. When she decided to run, my father's massive network plus my network were a good base for her. She was able to take the associations the family had and build a team from it."

It was certainly a family affair, and according to Bruce, Jim Clark was helpful in advising his daughter. "My dad told her, if she was serious about serving, she had to set aside her fears about money and a stable career and focus on the campaign. He was good at helping us focus, without telling us what to do."

Unfortunately, Jim Clark would not live to see his daughter win the election. He died of a sudden heart attack in 1995, two months after she won the nomination as candidate for Port Moody–Burnaby Mountain. Her father's death was devastating and served as something of a wake-up call. Her dad had been a smoker, and she was a smoker. She turned her formidable focus onto a new challenge: lifestyle change.

"After her dad died," says Marissen, "she wanted to clean herself up. She quit smoking and dyed her hair red instead of blonde. She started jogging, and she lost weight. CBC did a story on her as a rising star, all because she was getting into shape. This was very good for her political image, and ironically, this was when people started taking her seriously as a political player."

Derek Raymaker too noticed the changes she made. "She was a bit overweight, eating poorly, was smoking, and she decided to turn it around. She completely changed herself in less than a year. She was a woman in a man's world and was acutely aware of her body image being used against her. She is incredibly disciplined. When she sets her mind on a goal, she just puts everything into it."

Her campaign attracted new supporters. Vancouver lawyer Sharon White started volunteering to help Christy Clark after Bruce convinced her to attend an organizational meeting in Port Moody. "We had mutual friends," says White, "good friends who had attended SFU with [her]. All my friends were on the conservative side; she was on the liberal. When she wanted to run for the nomination, they lobbied me to come and help her. The first thing I remember is her brother Bruce recruiting me to a meeting out in Port Moody, and he said it was a twenty-minute drive. It was raining and dark, and I got lost. It was much farther than I had realized, and at one point I thought I was driving to the end of civilization. We still laugh about that. I worked on that nomination and helped her win. Her riding was a long way from my home turf, and way out of my comfort zone, but she was worth it."

White recalls that, in addition to the work, "We had a great deal of fun, as we always do. I remember going clothes shopping to get an outfit for the nomination meeting. For me, she has been inspiring."[1]

Despite McDonald's analysis of the riding, it was a tough fight, and a lot of observers did not think she would win. "It was an election against a strong NDP incumbent, very difficult. "I don't think Campbell expected her to pull it off. I think she knew she was going to do it, and it would take a year out of her life before the election was even called."

Mark Marissen moved to BC from Ottawa about a year after Christy Clark moved home. He was able to keep his job with David Anderson and transfer to the minister's regional office in Vancouver. He helped with Clark's campaign as her communications chair. He said there was no awkwardness about being the candidate's boyfriend and that they all worked very hard on the campaign. Clark was a natural with the media.

"She sent press releases to local papers every day," says Marissen. "We would package them up, and then she would drive straight to all of the newsrooms to make herself available for interviews. She gave good short snappy quotes; the media loved her."

Former party president Floyd Sully said her campaign stood out. "I attended a fundraising dinner that she had in the riding. I was surprised that someone as young as she was could draw such a crowd. It was a classy event. I was impressed."

Christy Clark took a strong position to close the Burrard Thermal Gas Plant. Gordon Campbell came out in support.

Says Marissen, "Taking on Burrard Thermal was the key issue. This position had huge support amongst people who would otherwise vote NDP, although it wasn't shut down."

Derek Raymaker had some pointed comments about the dynamic between Campbell, Clark, and Marissen. Clark and Marissen were, of course, a package deal.

"It didn't take Gordon Campbell too long to see Mark as a threat."

Says Raymaker, "I met Gordon Campbell when he came to Ottawa to recruit Christy to run and a few times during the 2001 campaign while I was helping Christy out. He came across as a complete ass."[2]

On May 28, 1996, Christy Clark pulled off a victory in her riding, although Gordon Campbell lost the election to NDP Leader Glen Clark (no relation), who pulled off a surprise victory, returning the NDP to a second term as government. The NDP had been behind in the polls prior to Glen Clark's election as party leader in February 1996.

Glen Clark had previously served in the Harcourt government as Minister of Finance and Minister of Employment and Investment. He was first elected MLA in 1986 and served in Opposition for the riding of Vancouver–Kingsway.

The election campaign had been hard fought provincially by the two main parties, plus the Reform Party and the Progressive Democratic Alliance (PDA). Jack Weisgerber, first elected as a Social Credit MLA, had become leader of the Reform Party in 1995.

In 1996, Weisgerber and Richard Neufeld, also previously a Social Credit MLA, were both re-elected, this time for the Reform Party, and Gordon Wilson was re-elected in Powell River–Sunshine Coast as leader of the PDA. Since neither party had four seats, which was the minimum for official party status, all three were considered independent MLAs. The BC Liberal Party supporters blamed vote splitting by these parties for Campbell's loss. Weisgerber resigned as Reform leader after the election, and Neufeld crossed to the BC Liberal Party in 1997. Wilson stayed as PDA member until 1999 when he joined Glen Clark's government as a cabinet minister.

Christy Clark won her seat by only about four hundred votes. She was the first member of her family to run for office and win.

Marissen and Clark took some time for their personal life, and were married in the summer of 1996. Athana Mentzelopoulos was a bridesmaid, and Derek Raymaker was Marissen's best man.

"I organized her wedding for her," says Mentzelopoulos, "and with my friend Maia, we did all the cooking. It was on the Gulf Islands. Her sister Jennifer was the maid of honour."

Derek Raymaker says there was an unexpected political tone to the wedding. "Campbell gave a speech at the wedding, and it was all about how great he was. He may have mentioned Mark and Christy once; the rest of the speech was a political stump speech. It was a few months after he lost the election to Glen Clark, so he was trying to be a bit emotional, but I lost respect for him there.

"Campbell's remarks at the wedding hit incredibly sour notes at many different moments. It should have been a breeze, especially for a seasoned politician. All he had to do was get up there and say nice things about the bride and groom. A low point was when he made a crack that he didn't trust Mark with the knife to cut the wedding cake because Mark would probably stab him in the back, like the federal Liberal he is, and try to advance Christy's career. I thought, *No wonder this guy lost the election.*"

The comment was interesting because Marissen was not really involved in provincial politics, other than to support his wife.

After leaving his job with David Anderson's office, Mark Marissen set himself up in private business as a communications consultant in 1998, and he made sure he kept his political focus on the federal scene so that he had the freedom to engage in politics without raising any potential for conflict of interest. He continued his heavy political involvement, primarily as a volunteer.

"Within two years," says Mark, "we had taken over the whole federal Liberal apparatus in BC with a slate of Young Libs, including Sharon Apsey from Victoria."

Marissen wanted to see the federal Liberal Party place more emphasis on membership and electing Liberal MPs. He felt it was too elitist and focused on sucking up to the eastern establishment and the party leader. This meant that most of the members were irrelevant.

He noticed that the Liberal Party of Canada seldom won seats in BC, even though it won majorities in the rest of Canada, and believed that BC's lack of MPs had created a dangerous culture in the federal Liberal Party in BC, where power was concentrated in unelected party members based on who they knew. He thought this was fundamentally anti-democratic.

Marissen felt the focus had to change so that BC federal Liberal members could plan and win some ridings, so they would have elected

Members of Parliament representing them in Ottawa. He organized a team that supported this view.

"We took on Ross Fitzpatrick who, as a private citizen, was way more powerful than elected ministers like David Anderson. We won a huge victory, with about three-quarters of the delegates supporting our slate."

Senator Ross Fitzpatrick was the well-respected, self-made business-man who built his award-winning Cedar Creek Winery in Kelowna as a hallmark of his work. Fitzpatrick had channelled his political energies in solidifying his long-time friend Jean Chrétien's political hold on BC's federal Liberal Party. At the time that Marissen took him on, Ross Fitzpatrick was a leader in the Senate. Mark Marissen's victory for his team would come with a high cost to him in the party.

When Marissen won his battle, the losers painted the party take-over as a fight between members who supported Paul Martin for leader instead of Jean Chrétien. Martin and Chrétien's supporters were often wrestling for power, and Marissen saw his new label as a "Martinite" ironic because, up to that point, he had been a very strong Chrétien supporter.

Given Ross Fitzpatrick's powerful position, his close relationship with leader Jean Chrétien, and the perception of Marissen's team's motives, the outcome of the battle tainted Marissen and his team's relationship with Chrétien and pushed Marissen from a hardcore Chrétienite to a Martin supporter. Later, Marissen ran Paul Martin's campaign in BC for party leadership and then ran the federal campaign in BC over two elections.

In everything that Marissen did in his interaction with the federal Liberal party, he was restrained because of his marriage to Christy Clark, who was a prominent MLA in Gordon Campbell's Opposition caucus shortly after the 1996 election. Marissen had to moderate his political choices and how people perceived him. Fortunately, he did well in con-sulting, and was able to channel his energy into work. Over the years, he sometimes had partners and staff, but usually he preferred to have the flexibility of a small operation.

The move from a prominent political position to a private support-ive position was a price that Marissen needed to pay once Clark joined Campbell's caucus; he was very proud of Christy Clark's win. "She was the only Liberal elected in that area, surrounded by NDP MLAs."

Her win for the Liberals, in a region dominated by the NDP, was significant for women in the BC Liberal Party as well. Campbell's party had already been receiving attention for its lack of strong female candi-dates. In a *Vancouver Sun* article, Vaughn Palmer had noted, a year before the election:

> Gordon Campbell's B.C. Liberal party is being questioned on the issue
> of its commitment to 'small l-liberalism,' this time over the dearth of
> women candidates running under the party banner.[3]

Notably, Palmer mentioned Clark as one of the candidates who could help rectify the image.

When one consents to marry, one consents to
be truly known, which is an ominous prospect.
LEON WIESELTIER

CHAPTER 10

THE PRESS GALLERY

EVERY ELECTED SEAT of government, whether the United State
Senate, the House of Commons in Ottawa, or the Legislative
Assembly in British Columbia, has a press gallery, which is the area
set aside for reporters whose job it is to report on the activities of government and its elected representatives.

The first official press gallery in the United States was the Reporters
Gallery for the Senate, created in 1841. Renowned writer Mark Twain
was a reporter in the Senate Press Gallery. He arrived in 1867 and, sometime thereafter, penned the following observation. "Senator: a person
who makes laws in Washington when not doing time."[1] This tone of coverage is often taken by reporters over 150 years later.

Reporters in the press gallery have the best seats in the legislature,
able to look down onto the floor of the House and observe the dynamics of the government and the Opposition members. For many reporters,
being assigned to the press gallery is the best possible assignment; you
have to earn the spot, and it comes with the responsibility of following
and reporting the actions of all the elected members, whether they want
you to report these or not.

In biology we learn about symbiosis, which basically means "living together," and is defined as "close and often long-term interaction

between two different biological species." I think most people who observe the press gallery and the elected Members of the Legislative Assembly will agree that there is a symbiotic relationship.

There is a mutually beneficial symbiosis between politicians and the media; they need each other, much like the clownfish and the anemone, where the anemone protects the fish from its predators, benefits by eating the clownfish's predators when they are lured close, and the clownfish fertilizes the anemone with its droppings, hanging out with the anemone because it has become immune to its poisonous sting.

In BC, the clownfish relationship is closest to that between the successful politicians and the press gallery. It is up to the reader to decide which, the politician or the reporter, is the clownfish, and which is the anemone.

The fact is, no politician can be successful without media coverage, and no press gallery reporter can do their job without material from politicians. For some politicians, it is an antagonistic relationship; for the most successful ones, it can be a relationship where both the media person and the politician achieve success at their jobs together.

The long-term leaders of the BC press gallery include Vaughn Palmer, Keith Baldrey, Mike Smyth, Les Leyne, and Justine Hunter. Vaughn Palmer is the provincial affairs political columnist for the *Vancouver Sun* newspaper, which dominates BC's print media along with the *Province*, and host of *Voice of BC*, the public affairs TV show on Shaw; Keith Baldrey is Global BC TV's chief political correspondent and a regional newspaper columnist; Mike Smyth is provincial affairs columnist for the *Province* newspaper and frequent radio host on CKNW; and Les Leyne is legislative bureau writer for the Victoria *Times Colonist*, the newspaper published in the provincial capital. Justine Hunter, the lone female veteran, writes for the *Globe and Mail*. It could be observed that Palmer, Baldrey, and Smyth are the political opinion triumvirate in BC, with a combined print and online reach for political reporting greater than any others, especially given their social media presence, and Les Leyne has huge influence in the capital city. These gentlemen have been in their various positions for some time, with Mike Smyth, the newcomer, starting in 1991; Vaughn Palmer, Les Leyne and Keith Baldrey full-time since 1984, 1985 and 1986, respectively. Christy Clark was nineteen years old when Vaughn Palmer began his job in the gallery.

Baldrey's station, Global TV, is arguably the single most influential news source in BC for decades, plus Baldrey pens a weekly news column for some community papers. Vaughn Palmer's column is well respected, and his TV program *Voice of BC* is influential and available to most BC residents on the Shaw network and online on Vimeo. Mike Smyth's column is widely read, and CKNW Radio, the political news station in Vancouver, has a wide audience. All this, taken together, gives you a sense for their collective influence and stature as opinion leaders in BC.

This book has benefited from the generous interviews provided by Keith Baldrey and Vaughn Palmer. I chased Mike Smyth for an interview for months, but he was uncharacteristically unavailable. I have considerable respect for the reporters in the press gallery, even though any review of the record will prove it is not mutual in terms of their regard for me.

In British Columbia we have a very testosterone-laden press gallery, which may be typical of all press galleries given that, until recently, politics was the domain of men.

Furthermore, the BC press gallery is also dominated by one demographic: men of "western" origin; in other words, a bunch of white guys. They are very qualified, highly informed white guys; however, given my particular perspective as an immigrant ethnic female, it has to be noted. This press gallery demographic is somewhat conspicuous given the increasingly diverse population in BC, including the largest Aboriginal population of any of the provinces, a growing, influential population from South Asia, and others from many parts of the globe.

Christy Clark, like other politicians, including former Social Credit cabinet minister Rafe Mair, former NDP cabinet minister Moe Sihota, and myself, moved seamlessly between politics and media. When Clark exited politics in 2006, she took over as a radio host on CKNW and was in that job as a talk-show host when she decided to re-enter politics in 2011.

Christy Clark's understanding of the job, on both sides of the microphone, has likely given her considerable strengths in responding to current events and ensuring that the message presented to the public is what she intends.

Public office in this province is not for the faint of heart. British Columbia is a vicious political battleground, and few who answer the call and step forward remain unscathed long. Frequently, blood is drawn by the interaction with the press gallery.

What greater thing is there for two human souls than to feel
they are joined for life—to strengthen each other in all labour,
to rest on each other in all sorrow, to minister to each other
in all pain, to be one with each other in silent unspeakable
memories at the moment of last parting.

GEORGE ELIOT, *Adam Bede*

CHAPTER 11

OPPOSITION MAGIC

B Y THE TIME the May 1996 general election occurred, Glen Clark
was the NDP premier, having replaced Mike Harcourt as leader in
February. Clark led his NDP to a surprising win against Campbell
with a leadership style described as populist. Just before the general
election, there were five political parties represented in the BC legislature, which was extremely odd for a province used to two big political
parties fighting for power, one on the right wing and one on the left. The
BC legislature included the governing NDP, the Opposition BC Liberals,
the BC Reform Party (originally Social Credit MLAs), the Progressive
Democratic Alliance (originally Liberal MLAs), and one Social Credit
MLA. This meant that, in the election campaign, there were five leaders
putting out policy and campaigning for votes: Premier Glen Clark, Opposition Leader Gordon Campbell, Reform Leader Jack Weisgerber, PDA
Leader Gordon Wilson, and Larry Gillanders, the unelected Social Credit
Party Leader.

Having a split in the so-called right-wing vote was a problem for Gordon Campbell; however, his party was far ahead in the polls, and it was
not seen as a big obstacle to Campbell's victory. There were some major
differences in the parties' platforms, however; and the leaders were
quick to point them out. One key difference between the party policy

of Weisgerber's BC Reform Party was the need to keep BC Rail a public asset, whereas Campbell's BC Liberals talked about selling it. Weisgerber, a populist leader from BC's northeast, put forward resource-based economic development issues that made Campbell's policies seem more urban. Wilson's policies went after both Clark and Campbell on a number of issues on the right and the left. Gillanders, also from the Interior, was flying the flag for Social Credit in the leaders' debates.

In late April, just before the election, Clark began aggressively questioning Campbell's platform for its planned cuts in spending while trying to bring support back to the NDP by promising to cut taxes. Les Leyne wrote that, in Glen Clark's TV address, which cost the NDP $50,000, he "painted a harsh contrast between his NDP government and the budget-slashing platform of B.C. Liberal Leader Gordon Campbell."[1]

In addition, Premier Glen Clark left behind his socialist image when he promised that small and new businesses would get a tax break and the existing freeze on tax increases would be extended by three years. This was connected to the NDP's promise to protect funding to health care and education with modest cuts, which Clark claimed provided a "clear signal of whose side my government is on." Glen Clark was filmed saying this at the school in East Vancouver that he attended as a child, invoking images of caring for the financially less advantaged voters from this part of the city. His slogan? *On Your Side.*

"Opposition parties slammed the speech as a 'deathbed repentance,'" wrote Leyne, "and a return to the days of 'us versus them' class warfare."

Glen Clark, only eight weeks into his term as premier, shifted from Harcourt's emphasis on a debt-management plan, in place for almost five years, toward tax cuts and program expenditures. Then he raised the idea that Campbell would cater to large corporations and turn away from government services. Leyne reported that Glen Clark "accused the former Vancouver mayor of wanting to give a $1.1-billion tax break to banks, developers and corporations" and "Campbell's plan to cut $3 billion from the $19-billion provincial budget would require the wholesale shutdown of many services."

Foreshadowing the division between right-wing Reform leader Jack Weisgerber and Gordon Campbell, Weisgerber provided this response to Glen Clark's televised address: "You could have substituted Gordon Campbell and one would have hardly noticed the difference." Weisgerber felt that Campbell's policies were too closely aligned with the NDP, especially in terms of First Nations issues. The BC Reform Party was strongly opposed to increased accommodation of Aboriginal rights and was opposed to treaty negotiations, believing that this would alienate the natural resources from general ownership.

In classic Glen Clark style, he wrapped up the tone of his pending campaign.

> Referring to criticism that the various goodies he has announced during the pre-election period are just political moves, Clark said: 'If helping you and your family instead of banks and big business is political, then I plead guilty.'[2]

In British Columbia, voters often react to the personalities of the leaders, and this reaction was observed by some reporters during the 1996 campaign. Mike Smyth wrote about the resurgence of the NDP after Clark took over as leader and noted Campbell made a misstep but was still expected to win the election.

> A current TV ad shows Liberal Leader Gordon Campbell talking about his usual issues—tax cuts, leaner government, debt reduction.
>
> The big difference, however, is what he's wearing. Gone is the business suit. In its place is a red plaid shirt that wouldn't look out of place in a loggers' bar.
>
> Political analysts have seized on the ad as an example of what's wrong with the Liberal leader, a former real-estate developer and Vancouver mayor regarded as a champion of the elite business class.
>
> With Premier Clark leading a revival of the governing NDP, analysts say Campbell is under the gun to soften his corporate image.

Unfortunately for Campbell, the TV makeover apparently hasn't caught on. The Liberals have seen their lead in opinion polls dwindle from 30 points ahead of the NDP to a neck-and-neck race in the build-up to a likely spring election.

Many analysts say Campbell's re-direction was an attempt to out-flank the Reform party and prevent a right-wing split that would let the NDP cling to power.

While Campbell has struggled, Clark has re-energized the NDP after its string of scandals and political gaffes.[3]

Long-time press gallery writer Jim Hume wrote a humorous perspective of the visual difference in the two main leaders in a *Times Colonist* piece entitled "Politics comes down to shirts":

Liberal leader Gordon Campbell is witnessed on television wearing a shirt of dubious plaid and trying hard to look like the kid from the farm down the road.

NDP leader Premier [Glen] Clark pops up daily on the same TV screen wearing a dark blue Howe Street suit with crisp shirt and tie and, more often than not, a fresh flower tucked in the button hole.

One of these two will be premier after election day dust clears, and I'm not sure which of them I prefer to be guiding my destiny: the businessman disguised as the "aw, shucks!" kid straight from the farm, or the blue-collar trade union organizer trying desperately hard to look like Daddy Warbucks fresh from the boardroom.[4]

As the polls showed the BC Liberal Party's lead shrinking, with the NDP gaining momentum, there were stories surfacing about pressure on BC Reform candidates to join the BC Liberals to prevent the NDP's re-election. Debates between the five leaders, although characterized by observers as not significant, showed a trend away from Campbell and toward Clark, with small nudges up in support for BC Reform, enough to cause problems for Campbell. The PDA, a new party, gained a bit of support; however, polling showed it was drawing from both the NDP and the Liberals, and support was to be nominal.

The BC Rail issue was key for many northern and Interior voters. One of the quotable quotes from the 1996 campaign was from PDA Leader Gordon Wilson, who said, "Our northern railway system is critical to the well-being of northern communities. The wealth of this great province comes from it regions."[5] Ironically, the biggest beneficiary of Wilson's vocal opposition to the plan to sell BC Rail was Weisgerber and BC Reform.

Then, on May 25, just a few days before election day, Social Credit Leader Larry Gillanders quit, and as he left politics he urged voters to vote for whichever candidate could defeat the NDP. As Dirk Meisner and Les Leyne reported in the *Times Colonist*:

> Gillanders, a real estate appraiser who has proudly declared that he was not a professional politician, instead urged British Columbians [to] vote for the best candidate who can defeat the NDP. "Four days before voting day, we stand on the precipice of an NDP re-election," he said. "That is because of the unwillingness of the other free-enterprise party leaders to demonstrate genuine leadership.[6]

Even as Gillanders resigned, he did not indicate support for Gordon Campbell. In addition, although he had resigned, the Social Credit Party still had candidates running in the election, although it was at about 1 percent in the opinion polls.

Two days later, Vaughn Palmer reported on Jack Weisgerber's claims that the BC Liberals were pressuring BC Reform candidates to quit or face consequences.

> Reform leader Jack Weisgerber claims the Liberals are subjecting his candidates to the carrot-and-stick treatment—"told they will be given certain advantages if they come over to the Liberals [or] the wrath of the business community will fall on them if they don't."

Still I'd be surprised if Campbell, his lieutenant Gary Farrell-Collins or any other top-rank Grits would take a direct hand in deal-making or intimidation . . . I would be equally surprised if some less highly placed Liberals were not up to their snouts in the shakedown. The temptation is too great, given opinion polls—today's numbers from Angus Reid are the latest—showing the two major parties neck-and-neck with Reform scooping more than enough votes to deny victory to the Opposition.

For the Liberals, these results are the fruits of a gross miscalculation.

When the Reform leader raised concerns about native land claims, the Liberals treated him like a rube from the sticks. Then later, when it turned out Weisgerber had tapped into a growing public backlash, they pinched enough of Reform B.C.'s platform to insult the man.

The promise to privatize B.C. Rail overlooked legitimate concerns in the ridings where Reform votes are concentrated. Now, having passed up every chance to court Weisgerber, the Liberals have no choice but to employ roughhouse tactics.[7]

Weisgerber, known as "Gentleman Jack" to political observers, expressed his anger that his party and his candidates were being treated poorly by the BC Liberals in a bid for power. These public expressions entrenched the urban–rural divide.

In the end, Glen Clark and the NDP were assisted in their re-election by a split in the right-wing vote between the BC Liberals and the BC Reform Party, and the public's perception of Gordon Campbell as working for the corporations versus Glen Clark working for the people.

The 1996 election returned Glen Clark as premier and Gordon Campbell as leader of the Official Opposition, which was not the outcome that BC's business community wanted; they wanted the NDP out. The right wing were not happy losers, especially given that the BC Liberals won 3 percent more of the popular vote than the NDP. It was the breakdown of the vote in the ridings that resulted in thirty-nine seats for the NDP, thirty-three for the Liberals, two for BC Reform, and one for the PDA.[8]

Glen Clark's win motivated the free-enterprise movement in BC to unite behind Campbell in order to ensure defeat of the NDP in the following election. Only days later, political commentator Rafe Mair wrote an article in the *Financial Post* calling on all the non-socialist political leaders to join with Gordon Campbell to make sure the NDP would be defeated in the future. Mair, a former Social Credit cabinet minister, noted that he had supported BC Reform as a possible replacement to Social Credit, but he had abandoned that in favour of the BC Liberals before the election, and he urged others to unite the right.[9]

Even though the NDP won the election, Christy Clark was elected in 1996 in an NDP riding as part of Campbell's expanded caucus. Campbell had taken the BC Liberals to thirty-three seats, an increase from the seventeen elected in 1991, and had grown the BC Liberal Party into the free-enterprise coalition. Campbell did benefit from a united right wing when Jack Weisgerber resigned as BC Reform leader in early 1997, and by November of 1997, both Weisgerber and Neufeld had resigned from the BC Reform Party.[10]

In Christy Clark's first term as MLA, she created a reputation as a fiery debater with a keen ability to articulate the views of the BC Liberal Party. She was immediately attracting media attention, even in the dog days of summer. She was the Official Opposition environment critic and took aim at any issue she felt was important. The first one was what she called a broken promise and, as Don Hauka of the *Province* reported, she obtained results:

Liberal MLA Christy Clark (Port Moody–Burnaby Mountain) released documents yesterday showing the ministry of finance promised on May 15 to pay 50 percent of the dental benefits for pensioners eligible under the Municipal Pension Plan.

But on June 18, three weeks after the election, the ministry sent another memo asking seniors to "please disregard and destroy" previous information on the dental-plan pledge.

Instead, retired municipal civil servants or their spouses will get a 20-percent subsidy. That means a single pensioner will pay $107 a year more than promised.

Finance Minister Andrew Petter promised to look into the matter.[11]

Her taste for the media and the role of Opposition was clear from the beginning, even though she was one of the rookie MLAs. As the NDP was trying to manage a complete overhaul of the forestry practices in BC, she was focused on a number of environmental issues. In July of 1996, she roasted the government in a story filed by Jim Beatty of the *Vancouver Sun*.

A forest company that felled trees in Garibaldi provincial park will face huge penalties if it is not more careful in future, Forests Minister David Zirnhelt warned Tuesday.

The infraction happened sometime between late 1994 and early 1995, but only became publicly known this week when the *Vancouver Sun* obtained documents through the Freedom of Information Act.

The company was fined $198,000 for the infraction—the maximum possible under the previous legislation, said Zirnhelt.

But Liberal environment critic Christy Clark said that when the value of the wood is subtracted from the fine, the penalty amounted to about $25,000. She called that a pittance in relation to the environmental damage and intrusion on fish habitat.[12]

She must have been a dynamo on issues because, at the same time she was going after the government on forestry issues and pension payments, she was following up on one of the issues that helped win her seat: air quality and the Burrard Thermal Plant.

B.C. Hydro's decision to fire up the Burrard Thermal generating plant is bad news for Fraser Valley residents already suffering from bad air quality, Liberal environment critic Christy Clark charged Tuesday in the B.C. legislature.

Clark wondered why Environment Minister Paul Ramsey would allow the plant to operate at a time when air quality is at its worst.

Ramsey responded that B.C. Hydro is already acting on a series of recommendations from the B.C. Utilities Commission to clean up the Port Moody plant, and that Clark is aware of the clean-up.

The plant became a source of controversy during the recent provincial election campaign after Liberal leader Gordon Campbell and Clark, now the MLA for Port Moody–Burnaby Mountain, portrayed it as a black mark against the NDP's environmental record.[13]

Clark was relentless and soon made a name for herself as one of the leading Opposition members. In an article for the *Province*, press gallery reporter Mike Smyth wrote:

There's nobody on the opposition benches who can drive an NDP cabinet minister into a trembling, voice-cracking rage quite like Christy Clark.

Not even Public Services Minister Moe Sihota—a guy rarely rattled by even the fiercest opposition baiting—could keep his cool yesterday once Clark began cranking up the heat.

Clark, who has quickly emerged as the Liberals' toughest, most relentless critic, deftly turned the proceedings into a battle of political lightsabres that would make Darth Maul proud.

She started by mocking Sihota's claim that the New Democrats are showing excellent leadership in government.

Then she started in on the fudge-it-budget scandal that erupted after the 1996 election.

Sihota, who normally swats away young opposition upstarts like so many gnats, just glared at her.

Then Clark zeroed in on the government's gaudy patronage record.

By the time Sihota got to his feet, he was furious... Sihota thundered, his voice breaking with anger... As Sihota flamed out, Clark just laughed at him, clearly delighted at getting such a rise out of the NDP's legendary pit bull.

This one was more for the entertainment value as Moe was bested by a rising young critic.[14]

Glen Clark's government had a very antagonistic relationship with Gordon Campbell's Liberals. Moe Sihota, the first Indo-Canadian elected in Canada and the first to serve in any cabinet, was a star minister and worked closely with Glen Clark in setting government policy and strategies. Sihota could see right away that Christy Clark had political talents.

"I first recall seeing her shortly after she was elected in 1996," says Sihota. "We formed government and I was education minister at the time, and she was my critic.

"She was part of the Liberal rat pack; she was very aggressive in Question Period. I thought the best way to start the relationship was to ask her to come meet with me. So she did."[15]

In the first meeting, Sihota remembers thinking that Christy Clark seemed unsure why a government minister would want to meet with an Opposition critic.

"I told her I wanted to know what her agenda was, so she could guide me in my thinking. It didn't seem like she had much of an agenda. It became apparent to me that she was not a policy wonk; she understood the game of politics, and she was eager to play the game. I don't remember having much of a direct dialogue after that."

If Christy Clark had a policy agenda, she didn't share it with the NDP. Says Sihota, "[Her agenda] was a lot like that of Glen Clark, Dan Miller and myself in 1988 when we were first elected, which was to go after the government. I don't remember her asking me much in terms of questions or on education policy, she went after the issue of the day. She was very much into the game of politics and not much substance."

Opposition members' job is to highlight where the government is tripping up in its mandate on behalf of the people to try to capture the attention of the media. Headlines are everything for Opposition members, something that Sihota understood well: "She was skillful at procuring a sound bite, she was comfortable in front of the camera, and she had one objective in mind, which was to help her party get to power."

Many observers comment on the gender dynamic, and whether it helps or hinders women in Question Period. Sihota says that while some women have a tough time and perhaps don't engage as aggressively as men in the bloodsport of BC politics, this was not the case with Christy Clark.

"[She] was a woman among men and understood the game of politics as well as the men did . . . There are people who just excel at the theatre of the legislature, and she was one of them. She loved it and she continues to love it."[16]

In many reviews of Christy Clark and Glen Clark, fans and opponents comment about the traits that both Clarks share in common. Keith Baldrey put this comparison together:

Clark and Clark:
- Good at tripping up their opponents
- Both are very likeable
- Similar in Opposition—very good at zeroing in on key points, making the points
- Both youthful, immature, when in government the errors you make become magnified
- Brash, creative, populist, both street fighters
- Glen Clark, like Christy Clark, could really get under the skin of his opponents

Sihota's observations are very similar. "An attribute of a good politician is the interaction with the public and how much they enjoy the engagement. Some people are very natural at it, and she's one of them. Glen Clark was the same way; they enjoy meeting people and people sense that."

Mark Marissen said their married life during this time was good, although very busy. "It was a frenetic life, with not a lot of down time. My phone was ringing 24/7, and so was hers, but mine probably more so. I was on a permanent campaign from about 1990 until 2011. I was always working on some kind of next election."

Christy Clark's childhood tradition of taking off blocks of time for family continued, even in a busy political marriage. "She would always take off significant time in August and Christmas and other times, and we would just shut everything off. We usually hosted the family dinners, we did a lot of hosting of Christmas dinners. There was a lot of entertaining in our household. In fact, more than she wanted. I tended to be driving that. We did have lots of fun parties. There were not many romantic dinners."

Meanwhile, there was a period of time when Christy Clark went after Gordon Wilson, and it became very scrappy.

First, to set up the context, Gordon Wilson had crossed over to the NDP in 1999, dissolving the PDA and moving into Glen Clark's cabinet as the Minister of Aboriginal Affairs and Minister Responsible for BC Ferries, taking on two hot potatoes at once: the Nisga'a Treaty ratification in the legislature, and the fallout from the Fast Ferries controversy. As the MLA for Powell River–Sunshine Coast, he often had to take floatplanes to access parts of his riding, and some of the rural parts of the province where he had meetings with First Nations as Minister of Aboriginal Affairs. Christy Clark had been going after Gordon Wilson in Question Period over his expenses for floatplanes, and it had become an almost daily exchange that was very heated, very noisy, and classic BC legislature antagonism. It all came to a head when Gordon responded with a memorable quote.

Marissen remembers it well, with a grin, "The Gordon Wilson moment was one of the most iconic moments of her career. I was home at the time, she had just gotten back from Victoria, was having a big spat with Gordon Wilson about how he was overspending by five thousand dollars because he was taking floatplanes rather than regularly scheduled flights. He said something about 'not having access to a broom,' basically calling her a witch. So a BCTV camera truck shows up in front of our house. She has a very serious look on her face as she goes outside to answer questions."

Marissen was watching everything nervously, having followed the controversy already, and not sure how the interview with BCTV (now Global BC TV) would play out. "As she is being interviewed, I see our neighbour's black cat wander across our yard toward her, and I think, *This is going be a disaster.*"

Marissen starts laughing loudly as he recalls the rest of the story. "Instead, she picks up the cat! She starts stroking this black cat, and it isn't even ours, and she just replies to Wilson's comments as if it is the most normal thing in the world. It was the top story on the news. BCTV even played the music from *Bewitched*! They played film of the cat rolling around on our lawn. She skewered him!"

Gordon Wilson has a slightly different view. "Because I represented a riding that was expensive to travel to and from, because of the lack of scheduled airlines and road connections, you had to charter flights. My expenses as MLA were generally higher than the other MLAs. She had gotten into the kick of going after my expenses. My mother, who used to watch Question Period, said, 'Next time she jumps up to do this, tell her the reason your expenses are so high is because you don't have a broom.'"

It was a funny comment, but in order for him to use it, the timing had to be right and Christy Clark would have to stand up in Question Period and attack him again. She did.

"Next time she stood up to come after me I stood up and said, 'Well, unlike the member opposite, I don't have a broom to ride around on.' And that phrase was a good shot. It stopped her and she seemed furious."

But Wilson paid a price for using a comment like that while sitting as an NDP cabinet minister. "In the scrum afterwards there was rising indignation that I could use such a term. I thought it was humorous.

Even then I admired Christy Clark as a parliamentarian. I took a huge amount of flack from the women in the NDP caucus who thought the comment was completely inappropriate."

Lee Mackenzie, CHEK TV noon news anchor, offers her perspective. Mackenzie was in the interesting position of being both a member of the media and a friend of Gordon Wilson, who was at her family home on the evening of the debate. Mackenzie says, "I remember there was heated debate between Gordon Wilson and Christy Clark over the subject of some of his expenses for flights. She was feisty and fearless. When she stood up to speak you could just feel her energy fly across the room. She appeared to love a pitched battle of words and relished the chance to get up and start swinging. There was a serious edge to her attacks. Although Gordon didn't seem to in any way be beaten or battered by her onslaught, he did launch a volley of his own.

"As a member of the news media, I remember the reaction from myself and others—a collective gasp followed by the urge to laugh out loud. It was a cheeky, clever, and daring line, delivered with great skill. Vintage Wilson. It only served to throw fuel on Christy Clark's fire, but I have no doubt whatsoever that she loved it."

What about the issue itself of overspending on travel? Was the focus effective?

Says Wilson, "Her questions when she went after my travel expenses were ridiculous. Every member from this riding is vulnerable to those questions because there's no other way around. The questions were designed to defame me because here's the guy who was the leader who brought her party into Opposition and he's sitting with the enemy, so she wanted to eviscerate me. I don't think she hated me. My guess is she was angry with me. We had shared a vision, there were a group of us, we were idealistic, and she was still trying to realize it."

BC politics was pitting two former allies against each other in a high-profile way, with a rising star in Campbell's caucus going after a prominent NDP cabinet minister, her former leader. "I think that what they were trying to do was destroy my credibility so that I had limited prospect in the NDP, so that I could never, for example, become leader of the NDP. There's no question from a political strategic point of view that was the goal."

Wilson admits she was effective at getting under his skin. "On the one hand, it drove me crazy. On the other hand, I thought if they are investing that much time in attacking me, I must be doing something right.

"Christy Clark was a pit bull," Wilson says. "In the Campbell Liberals she was probably one of the most feared members of the Opposition. She could think on her feet, she had a sharp tongue, and she drove to the heart of the issue. Because she was effective, she was disliked by many members of the NDP, especially the women."

The comment that women disliked her more than men did was one that would come up later in Clark's political career. I asked Wilson about this. "I found it interesting that Premier Glen Clark seemed to have more respect and fun with her in debate than the female ministers did. It seemed there might be a certain amount of jealousy or envy. She was an attractive woman who didn't try to hide it."

Whereas she might have been annoying Sihota and Wilson with her effectiveness in Opposition, she was winning over fans in Campbell's BC Liberal Party. One person who joined the BC Liberal Party after Campbell became leader was Brad Bennett, whose grandfather, W.A.C. Bennett, holds the record for longest-serving premier of BC, and whose father, Bill Bennett, served as a popular Social Credit premier. Brad Bennett was asked to be involved at the organizational level of the party, particularly with helping with provincial conventions. While doing these jobs, he met Christy Clark.

Conventions are all about networking, especially for emerging political players. One of the key networking opportunities is a hospitality suite, which is a room set up for a socializing, often with drinks, snacks and music. Word always gets around conventions about who has the best hospitality suite.

Of Christy Clark, Bennett says, "She always had the best location and biggest signage for the hospitality suites. Her rooms were always a real standout, and by everyone's account she always had the best hospitality suites at conventions. Lots of fun."

When Brad Bennett had a chance to get to know Clark, he found her appealing. "She's a real people person. I think that's a big part of her success as a politician, her ability to connect with people in a very real and honest way. People can spot insincerity a mile away. With her, what you see is what you get. She speaks from the heart. I think that resonates."

Bennett had no interest in following in his father and grandfather's footsteps and largely avoided politics until he re-engaged with the BC Liberal Party under Campbell. He had to admit, however, that his encounters with Clark always impressed and motivated him.

He appreciated "her high intellect, and quick wit. She has a great sense of humour; it's the whole package."

The crucial flaw in his old life was the belief that he could be the driver of his own journey. So long as you believe that you are the captain of your life, you will be drifting farther and farther from the truth.

DAVID BROOKS

CHAPTER 12

CAMPBELL'S CABINET

B ETWEEN THE 1996 and 2001 elections, the NDP went through considerable turmoil and change, while the BC Liberal Party consolidated its place as the new free-enterprise coalition, effectively rebuilding the Social Credit base under the leadership of Gordon Campbell.

Glen Clark was forced to resign over an allegation that he received discounted deck construction on his home in exchange for consideration of a neighbour's casino licence application. Although Glen Clark was later found innocent, the investigation initiated by Attorney General Ujjal Dosanjh ended Clark's political career. In the following leadership race, Dosanjh won the leadership over Corky Evans, the well-liked Kootenay parliamentarian who served in Glen Clark's cabinet until 1999.

In the 2001 election, BC was again a province with two main political parties, Gordon Campbell's BC Liberals and the NDP with Premier Ujjal Dosanjh. The Green Party, under leader Adriane Carr, ran a high-profile campaign, with Carr joining the televised leaders' debate. She ran in a three-way race in Powell River–Sunshine Coast against Gordon Wilson and Liberal Harold Long, a former Socred MLA. Long was victorious in that race, and the Green Party failed to win any seats in the election.

The May 16, 2001, election proved disastrous for the NDP, reducing them from government to only two seats: Vancouver-based MLAs Joy MacPhail and Jenny Kwan. Gordon Campbell won the most decisive general-election victory in BC history when the Liberals won seventy-seven out of the seventy-nine seats. Christy Clark was one of the MLAs elected and soon became one of Campbell's most prominent cabinet ministers.

Christy Clark was pregnant during the election, and not just *a little pregnant*; she won her riding in late May and was due less than three months later. This was her first pregnancy, and she did not slow down.

What was remarkable was that her pregnancy was not a negative factor in her campaign, and in fact, appeared to be a huge political asset, according to Marissen. The 2001 election was only ten years after I had run for office, and I was not able to publicly speak about the fact that I was pregnant because at that time it would likely have been a barrier to my election. By 2001, the idea was accepted, especially since Clark was clearly a young, energetic, bright candidate who was prepared to manage her family situation as needed.

Given that Campbell had seventy-seven MLAs from which to choose, he could have easily justified leaving Clark on the backbench while she managed her first experience with motherhood. Instead, she was front and centre as Minister of Education and deputy premier when Campbell named his cabinet on June 6. It was to Clark's credit that no one questioned her ability to take on these jobs while welcoming her baby. On the contrary, Clark's pregnancy was woven into political commentary as if it was the most natural thing in the world. Press gallery reporter Mike Smyth wrote:

Take a good look at Christy Clark and Gordon Campbell and one thing's obvious:

They're both due. Big time.

Clark, the dynamic education minister, is due to deliver her first child any day now.

She is so close to the big day that the government's education bills have been fast-tracked in the legislature. Last week's Bill 8, enshrining parents' rights to volunteer in schools, was in the nick of time.[1]

The portfolio of education was expected to be difficult, because Campbell's government was at odds with the BC Teachers' Federation on many contract issues, and Clark's term as Minister of Education is the one people refer to as Clark's "war on teachers." The stage was set for a showdown with labour, as reporter Lori Culbert wrote in the *Vancouver Sun*:

> The situation with teachers, who enter contract negotiations in the fall, is one of four major labour challenges the new government faces. But Clark, who is also deputy premier, would not comment on the possibility of a labour crisis brewing in the province.
>
> "I can't predict if that is going to happen in the future. I do know, though, that we have a mandate from British Columbians to keep these promises and making education an essential service is something that is going to happen in the first 90 days of this government," she said, after the new Liberal cabinet was sworn in Tuesday.[2]

Before the baby was born, there was a story written in the *Globe and Mail*'s Inside Politics that gave a clear view of how important both parents were in Canada's political landscape. Entitled "A most powerful baby," the story featured a large photo of Christy Clark smiling and very pregnant. It stated that, in addition to the baby's mother being the new BC government's thirty-five-year-old deputy premier and education minister, the baby's father is a "key regional operative in federal Finance Minister Paul Martin's leadership campaign."

Clark is quoted in the story, acknowledging that her friends think she is crazy. Calling her baby a "power baby," the story describes Clark and Marissen's plans after Clark's month off, which included sharing the costs of a nanny with another woman in the legislature with a child, accommodated in a small room in the legislative buildings. In the article, Clark is quoted as saying, "I have two choices—I can say, 'Oh, well, it's never been done,' or I can try to do it and make it work. I want to have kids and I want to have a job. I think you can integrate the two."[3] And she did, and became one of the first cabinet ministers in Canada to do so, setting a precedent for women in the future who sought to combine motherhood with a career in politics.

Christy Clark, as education minister, changed policies to help working mothers and moved forward with an agenda to change the education system to expand the role of parents and introduce a rating system, which was not something welcomed by the BCTF. As the *Times Colonist* reported on September 26, 2001:

> Parents and students will be invited to rate the B.C. education system once a year as part of a dramatic shake-up promised by the provincial government, Education Minister Christy Clark said Tuesday.
>
> Clark, who brought her baby son Hamish to work after a month off on maternity leave, said her ministry will solicit parent and student views... on a variety of academic and non-academic issues, including school safety.
>
> Clark said the surveys will signal to parents that their input is important and will eventually boost confidence in public schools... In designing the surveys, the minister said she will work closely with the B.C. Confederation of Parent Advisory Councils, the umbrella organization for local parent councils in schools around the province.
>
> Confederation president Reggi Balabanov said her group is excited about the prospect of parents and students having such a strong voice in educational reform, noting they can often sound alarm bells before school staff become aware of problems.
>
> Meanwhile, Clark urged employees to follow her lead and bring their babies to work to encourage employers to set up child-care centres.
>
> However, after their election in May, the Liberals scrapped a before- and after-school child-care plan introduced by the former NDP

government and eliminated NDP plans to progress to universal child care. The cost of that plan was not sustainable for government and did not address the problem of a lack of day-care spaces, Clark said.[4]

Clark had an ambitious agenda to manage immediately after her swearing in as Minister of Education, and she started work right away by introducing legislation that protected the rights of parents to volunteer in schools, while protecting existing jobs. She was forced to do this because there were rumblings that CUPE, the union overseeing non-teaching school employees, was moving to ban parents from volunteering in schools.

> "The point of this legislation is to make sure that no school district can ever permanently bargain away the right of a parent to volunteer at his or her school," Education Minister Christy Clark said Thursday.
>
> The legislation will also prohibit school boards from using volunteers to displace employees. That provision would be in play during any strike, Clark said.
>
> "The legislation is explicit. You will not be able to use volunteers to displace existing employees."
>
> The government is still promising to introduce legislation in the next few weeks that will make education an essential service, but Clark said the same rules on the use of volunteers will apply.[5]

That was the legislation Clark introduced in early August, and then she was in the countdown to her due date with her baby. Stories anticipating the baby revealed Clark's determination to try to make it work. In an article for the *Calgary Herald*, Lori Culbert wrote about Clark's level of preparation leading up to her planned month-long leave of absence. As the article winds down one gets a sense for the juggling act she has ahead of her.

> "One doesn't really know what it feels like not to be able to pull yourself out of a car by yourself until you actually confront the situation.

Or to have swollen ankles, or those kinds of things that every pregnant woman goes through," laughed Clark.

"I feel ripe. I feel very heavy and very full. But on the other hand, you're sustained by the pure joy and excitement of the fact that you're going to have a baby. The payoffs are so big."

Even during her month at home with the baby, Clark vows to be in communication with Victoria and her Port Moody constituency during that time to get her political fix.

"I'll have my e-mail on, I'll be in contact," says Clark, who is serving her second term in office and has worked in politics all her career.

"I'm going to stay in touch, but I don't intend to be diverted from my son when he's born . . . I know that when the baby comes, I am going to want to focus on enjoying my baby for a little while. And I'm going to want to give all my time to him."[6]

Clark integrated her new role as mother into her job as a powerful political figure; however, media and political observers have questioned how much power she truly had as a cabinet minister in the Campbell government. Keith Baldrey talks about the BC Teachers' Federation's contract negotiations.

"[Clark was] the one in the job when the class size language was stripped out of the contract. My view is that Gordon Campbell really ran almost every portfolio and that most ideas flowed directly from his office. I think he was more the education minister than Clark was. You could sort of see that in those so-called "open" cabinet meetings, when he liked to outtalk every one of his ministers to show he knew more about their portfolios than they did."

Vancouver Sun press gallery reporter Vaughn Palmer shares this assessment of Clark's time as Minister of Education. "Hard to point to any accomplishments other than losing a high profile struggle with the BCTF. One reason why I had my doubts about her abilities in government. As for deliberately antagonizing the BCTF, the contract-breaking originated with Campbell/Collins, not Clark. It was a budget issue, not one of education policy. She had to carry out the policy, but I don't think it was her idea."

Palmer thinks that Clark's time as minister in Campbell's government likely served her well later, when she had to resolve a BC Teachers' Federation contract as premier and she insisted that both parties bargain, refusing to impose a contract. "I do think she learned from the experience to ensure that the BCTF did have an exit strategy in the recent dispute and doubtless took some satisfaction that the outcome was so much better."

Brad Bennett has a different perspective on Christy Clark's time as Minister of Education. He believes that Clark wanted to change the power dynamic in public education. In his opinion, Clark was "trying to separate organized public labour from the education agenda, to try to separate the BC Teachers' Federation's control of the classrooms, to try to return some of the power to the teachers over the classroom decisions. Those efforts are ongoing. I applaud her for trying to do that for all the right reasons."

Political observer and wildlife consultant Guy Monty provides an outsider's view that reflects a lot of the commentary on social media. "I remember when she was in the Opposition as environment critic and recall watching her slamming the NDP on issues. This was the first time I was aware of her, in the late nineties. Then after Campbell won, I started to pay attention to her as Minister of Education. That's where my non-appreciation of her began. At the time, the media was still spending time reporting on what the government was doing; there was a lot of talk about how she was handling education: problems with the teachers' union and cuts to funding, and she was central to a lot of those discussions."

Monty, as a wildlife consultant, works with many scientists in environmental assessment and resource management planning. He is not a member of any political party and is one of those rare political observers who watches the televised legislative debates. He was one of the few detractors of the premier prepared to talk on the record about Premier Clark.

"In our house, and the general culture that I live in, a lot of things about Christy Clark and what she is disliked for has more to do with her years with Gordon Campbell and his government's record. Even though it became obvious that no one had a lot of say in how Campbell directed his government, she was generally despised for how she handled education. When the court case came out, saying that the way she dealt with the teachers' union was unconstitutional, that underscored how badly it had been handled. I think that came out around the time that she was re-entering politics, which solidified the opposition to her. This anger always simmers, and as other issues arise, it comes back as a general negative feeling against her."[7]

The gap in perspective on Christy Clark on the issue of education is considerable. Brad Bennett, who usually kept his opinions to himself, shares his opinion on this topic.

"I disagree vehemently with anyone who tries to say she dislikes teachers. Her dad was a teacher, she has a great amount of respect for teachers, and she believes in the rights and abilities of the individual over the broader collective. She thinks the teacher in the class will know best."

As for the BC Teachers' Federation's perspective that Clark ignores what the teachers want, Bennett adds, "It is too easy for the BC Teachers' Federation to say, 'This is what teachers say,' but I don't buy it. The BCTF is a big public sector union that wants to control the education curriculum and how we teach our kids. [Clark] really wants to try to do the right thing for teachers, students, for families and for taxpayers."

Gordon Wilson served as Minister of Education while in cabinet in the NDP government. He comments on the similarity of approach to the BCTF between the NDP and the Campbell government. "The education policy was fractious. As Minister of Education, she ended up carrying on the relationship with the BCTF that was started by the NDP, and that is legislating people back to work. The teachers were always considered an adversary by government, and this adversarial relationship carried on."

But how much power did ministers have in the Campbell government? According to *Vancouver Sun* reporter Vaughn Palmer, not much. "Campbell ran one of the most centralized premier's offices we have ever seen in Victoria, and that is saying something."

Keith Baldrey claims Campbell had "a 'control freak' approach to the job, where he didn't allow his ministers, particularly his younger ones, any latitude. (Clark) was very young."

Baldrey thinks that gender was a large factor in the working relationship between Campbell and strong women. "I think Clark chafed under Campbell's leadership," says Baldrey. "Other notable women also seemed unhappy with Campbell's style of leadership, such as Carole Taylor and Olga Ilich, Sindi Hawkins. A number of women had problems with Campbell. None of the strong women were particular fans of his. He ran a real 'testosterone government,' really masculine approach, men in charge."

Derek Raymaker is more blunt in his assessment of the dynamic.

"Campbell is such a sexist twit."

In addition to being a cabinet minister in the Campbell government, Christy Clark was made deputy premier, a role and title that sometimes carries considerable weight. When asked about Clark in this role, Palmer says, "Can't fairly judge her on that point because Gordon Campbell was too controlling to ever take the title seriously."

Baldrey agrees. "It's largely a gimmicky title, and I don't think it really meant much."

Gordon Wilson says the title of deputy premier was useful in terms of political strategy.

"I saw Campbell as a premier who controlled everything from the premier's office. I was surprised when he made her deputy premier because I didn't think that he would put a woman in that position of relative authority. It told me that he did that for one reason only: to keep the Liberals in the coalition that he needed to govern."

Guy Monty has a blunt assessment of the change in governing style that BC encountered under Gordon Campbell. "Campbell did what

Campbell wanted with no possibility for compromise or discussion at all."

Monty says this had a profound impact on communities and public engagement in decision making, especially for activists who wanted to contribute on issues like social justice or the environment.

"So many people that I knew who were active in their communities just gave up." Monty believes that the lack of access to information also impacted reporting and media stories. "The media gave up investigating what the government was up to, reporting what the government was up to, analyzing what the government was up to. You look at the total failure of the Freedom of Information office, making it harder and harder for anyone to know what the government is doing."

Guy Monty has been working in BC for decades and is passionate in his efforts to protect wildlife. He has worked on countless projects that require detailed research into wildlife inventories in order to ensure compliance with provincial and federal regulations and legislation. He says that the relationship that researchers like himself had with the BC civil service changed after the Campbell government was elected.

"I remember the first time someone said to me, 'The NDP lost; go somewhere else.' This was around 2003. I was trying to report a poacher, and the conservation officer tried to explain to me that there were no longer resources to report this. I was frustrated with him, he was frustrated with me, and finally he said to me, 'Look buddy, the NDP lost, and things are done differently now.'"

When Monty recounted this, his voice changed and he sounded defeated all over again, even as he recounted the story to me twelve years later. Monty's voice picked up passion, however, as he finished the story that for him was a turning point.

"I was absolutely shocked. I have worked in this field since the Social Credit Party was in power, and through the NDP years, and then when Campbell's Liberals came in, the change was drastic. There were no changes to the laws protecting wildlife in any visible sense, but the ability to enforce the laws was basically gone, through lack of resources to investigate, lack of courtroom space to try cases, and a policy that said if you were not 100 percent positive that you would be generating revenues (through fines) for the province, you are not to investigate."

At the time he told his story, in the fall of 2015, the headlines were about conservation officers overworked with wildlife issues involving black bears and wolves. Someone working in the field would understand the background to these headlines that might not be obvious to the rest of the province.

To Monty, the policy changes were a blatant conflict between business interests and progressive interests, and what was worse, it was being waged in settings that were supposed to be non-partisan and beyond politics.

"In the first five years after Campbell came into power, many of us who were progressive individuals trying to get information from the government, or even from a local agency like the chamber of commerce, would be shouted down or provided a political answer like 'the NDP lost.' The funny thing is, many of us were not NDP. We were just trying to participate with information."

If that was the perception for some outside the political process, the reverse was occurring on an individual basis for many people who had not been involved with government previously, and reviews from those who worked with Christy Clark directly were positive.

Multicultural strategist Jatinder Rai met Clark when she was in cabinet in Campbell's government. "I got to know her through Mark and met her in 2003. What I recollect of this meeting is that, after knowing Mark through federal activities, I reached out to Christy Clark when she was Minister of Education. She didn't know me from a hole in the wall, but she made time and met me for lunch to talk issues. I reached out to her because she seemed to me to be a person who genuinely connected with the South Asian community. There were a number of people who were politically connected, but they weren't necessarily that caring. She had close friends who were South Asian, and she seemed progressive, opinionated, and candid in her connection to the community."[8]

Although Rai was only marginally involved in politics, he had a long-standing interest in government and public policy. He felt that, after the Campbell government was elected in 2001, there was a chance to be involved. "Our conversation over lunch was wide ranging and very interesting. I wasn't looking for work or a job or anything personal, I was looking for a way to contribute to the process of public policy development. I found that when it came to important issues that matter to South Asians like community, family, and society, she was really grounded in the solutions; she was progressive."

Clark impressed Rai with her knowledge. "She was from the mainstream community, yet she understood the South Asians as well as an ethnic MLA did." He was also happy that she didn't identify with him based on his ethnicity, but just as a fellow human.

"I see myself as a British Columbian; however, I know I have diverse ethnic views, but these don't represent all of me, they make up small pieces of me, and when people only identify me through my ethnicity, they miss who I really am. She makes me feel accepted for who I am."

Long-time civil servant Dan Doyle served a number of governments in senior positions, always with exemplary reports of his work. He first met Christy Clark when she was Minister of Education in Campbell cabinet. He remembers a meeting he had with her as Minister of Education. It was an issue around school buses.[9]

"She felt really strongly. I was deputy minister of transportation. I told her at the time that I didn't think what was being proposed would work. She asked me why, and she listened and changed her mind. The key to this is she made the decision based on facts, not based on hearsay."

Doyle was impressed with Clark's intelligence. "It amazes me how quickly she can assimilate and understand issues and come to a decision."

This style of leadership is one that people who work closely with her frequently comment on: her ability to absorb information and change her opinion if presented with a series of compelling facts from an informed person.

In January of 2004, before the spring legislative session, Gordon Campbell shuffled his cabinet, and Christy Clark ended up in the unenviable role of Minister of Children and Families.[10]

Global BC TV press gallery reporter Keith Baldrey reconnected with Christy Clark after she was elected to government and remembers when, as a cabinet minister, she had to revisit some of her criticisms of the government that she had made while in Opposition.

"She was an effective critic while in Opposition and honed her communication skills rather well. I recall her being the critic for children and families, which eventually became a source of irony when she became the cabinet minister with that portfolio, as she then had to field some of the same over-the-top questions she threw in Opposition. I had this conversation with her, when she was complaining about some of the questions from both the media and the Opposition, and I said these are similar to what you did in Opposition. What goes around comes around. She laughed and admitted that she overdid it sometimes in Opposition."

"The ministry of children and family services was not a happy place—and this was one reason she left politics. The only time you are in front of the public is when you are dealing with a child who has died in care," observed Derek Raymaker. "Campbell probably thought it was a woman's portfolio and she fit the demographic profile of who the minister should be."

Keith Baldrey said the appointment was not welcomed by Clark.

"... she balked when Campbell told her he was moving her to the thankless children and families portfolio. The cabinet shuffle was started that particular day, as ministers were informed of their new assignments, but it suddenly ground to a halt when Clark at first said no to the switch. Campbell had to talk her into it."

Vaughn Palmer believed that this new portfolio was a key reason why Clark decided she did not want to run in the 2006 election.

Premier Gordon Campbell was in Ottawa Thursday morning, basking in the relative success of the first ministers' conference on health care.

Then Campbell got the surprise call from Christy Clark. The deputy premier and minister of children and family development had decided not to run in the next provincial election.

Campbell knew that when he got off the plane from Ottawa, the first questions would be about the untimely departure of the most prominent woman in his administration.

Clark joked Thursday that "I'm doing this so you'll stop calling me ambitious."

It was a reference to the way that I've highlighted her political drive and not the hankerings of some of her colleagues.

Not to disappoint Clark. But I think this decision can also be understood in terms of her political ambitions.

In this same article, Palmer goes on to talk about the conspicuous disconnect between her importance as a highly competent, prominent woman in a "government with a gender imbalance at the cabinet table and a sometimes-threatening gender gap in the polls" and the fact that she'd been moved to a "second-echelon ministry with a no-win mandate," one which Palmer characterizes as a "graveyard of political ambitions."

In short, I'm thinking Clark saw herself going nowhere in this government.

Which is not to dismiss her stated reason for deciding to step away from the cabinet table and give up her seat at the next election.

"Hamish is not going to be able to find another mom," she said. "The government is going to be able to find another politician."

Not likely. Her departure creates a hard-to-fill vacancy.

Campbell has a poor record of recruiting high-profile women to run for office and a worse record of promoting them when they get there.

I expect many would say that if the hard-driving Christy Clark couldn't hack it in this administration, neither could they.[11]

Mike Smyth of the *Province* had a slightly different take on things and ended his story about Clark's announcement with a prediction that was amazingly insightful:

I have little doubt Christy Clark wants to spend more time with Hamish, her three-year-old toddler, but it's a little difficult to accept that's the whole story behind her sudden resignation.

In fact, she and the kid milked the whole mother–baby political feel-good story for all it was worth. There were so many Christy–Hamish photo-ops and convention appearances that caucus colleagues jokingly referred to Hamish as Clark's press agent.

And now she wants to break up the act? No, there's something else going on here. And you don't have to dig very deep to come up with plausible theories.

He suggests that, while Clark's decision might be linked to fallout from the BC Rail investigation (discussed in Chapter 13), it certainly had a lot to do with Clark's shuffle into the no-win children and families ministry—a move tied to Clark's growing power base in the BC Liberals.

Clark had been one of the most dynamic (if controversial) education ministers the province had ever seen. But Campbell dumped her into the cabinet graveyard of the Children and Families Ministry, the worst career-killer in government.

Why did Campbell do it? Probably because he knew Christy was getting too popular and posed a possible leadership rival down the road . . . It could not have escaped Campbell's notice that Clark was the most popular cabinet minister in the Liberal Party.

Her resignation looks bad on Campbell, already very light on female talent in his cabinet and unpopular with women voters. (The timing of the announcement could not have pleased the premier, ensuring he was gooned by TV cameras at the airport as he returned from Ottawa.)

Something tells me Christy will be back—after Campbell is gone.[12]

We are all stumblers, and the beauty and meaning of life
are in the stumbling—in recognizing the stumbling and
trying to become more graceful as the years go by.
DAVID BROOKS

CHAPTER 13

BC RAIL

I N 2003, THE BC Rail controversy was a national story for Christy Clark,
and the way it was reported cast her and her husband, Mark Marissen,
plus her brother Bruce Clark, in a suspicious light. Some people believe,
based on what they have read or heard in the news, that Christy Clark
was involved in a corrupt giveaway of BC Rail and that her husband and
brother were involved, and they have all somehow managed to get away
with it.

When I asked him what people think of when they think of Premier
Christy Clark, wildlife consultant Guy Monty expresses a comment I
hear frequently. "People think of BC Rail, and I know that has dropped
out of media consciousness, but I think a lot of people associate her with
the BC Rail scandal."

When I began to write this book, I spent considerable time researching the BC Rail story because, even though I had heard about it for years, I
was never able to get a good sense for anything that felt like a solid story.
The funny thing is, the more research I did, the less convinced I was that
the story, as it was reported, was real at all, though I did find a pattern to
the names repeated as "central" to the story.

I could not find any real evidence linking the reported "bad guys"
to the BC Rail sale. Meanwhile, there were some names that were

seldom in the news even though they were key people in government. Premier Campbell's chief of staff, Martyn Brown, for example, was rarely mentioned, even though evidence in court shows he was heavily engaged in the file, acting in his capacity as chief of staff for Premier Gordon Campbell. And conspicuously, Campbell was not in the story at all.

The story broke in December of 2003, with images of a police raid on the BC legislature and allegations of corruption at the highest levels of government with respect to the sale of BC Rail. The government names in the story were cabinet ministers Gary Collins and Judith Reid, and their aides David Basi and Bob Virk. Excerpts from a story filed by Jim Beatty in the *Vancouver Sun* are clear about the connections of the key people at the beginning of the controversy.

On Monday, the premier's office fired Dave Basi, the top political aide to Finance Minister Gary Collins and suspended with pay Bob Virk, the top political assistant to Transportation Minister Judith Reid.

Both their legislature offices were raided Sunday by police, who left with dozens of boxes and a number of computer hard drives.

No one would say what specific allegations relate to Basi or Virk—only that they are related to a comprehensive, 20-month police investigation into commercial crime, drugs and organized crime.

Both Collins and Campbell dealt with the suggestion that the investigation into Basi and Virk may have something to do with the government's decision to privatize BC Rail.

Virk in the transportation ministry and Basi in the finance ministry were key players in the government's plans to privatize BC Rail earlier this month.

Fuelling speculation that the BC Rail privatization may have something to do with the investigation, a consulting company that was lobbying for one of the unsuccessful bidders was also raided.

Campbell said he has heard no information suggesting government policies or decisions were in any way compromised.

Collins remained confident in the decision to sell BC Rail to CN Rail and said an independent auditor is currently studying the deal and how it was reached.

The investigation extended as well to the husband of B.C.'s deputy premier.

Christy Clark said Monday she and her husband Mark Marissen first learned about the investigation Sunday when RCMP showed up to ask questions.

"They asked for his help," Clark said. "They felt he may have been the innocent recipient of correspondence or documents and of course, he cooperated fully."

While her husband talked to the police Sunday, Clark said she, herself, has not been questioned or interviewed.[1]

Very soon, though, the names that kept coming up concerning BC Rail were Christy Clark, her brother Bruce Clark, and her then-husband Mark Marissen. The stories made it sound like somehow the deputy premier had allowed her family members to work behind the scenes to benefit from the sale of the railway. When I began research into the story, I expected to find out why these were the three names most often mentioned, but instead I ended up somewhat confused.

For what it is worth, when I asked Christy Clark about how her name got associated with the BC Rail story, she seemed as baffled as I was.

"I have no idea."

When I called Bruce Clark to interview him for the book, I said I wanted to ask him about BC Rail. He replied that he had never done an interview about BC Rail and that he never would. I told him I only had one question:

"Why is your name even associated with the BC Rail controversy? I have looked everywhere and I cannot find a single connection."

He just laughed. "That is a very good question."

I will admit that I was surprised that I could not to find any connection between Christy Clark and the BC Rail scandal, because I had heard so many allegations over the years that I expected to find *something*.

So whose names were central in the story that made so many national headlines? This story kept popping up, even as recently as the 2013 general election, suggesting corruption connected to Christy Clark. The names repeated were Erik Bornmann, Jamie Elmhirst, Bruce Clark, and Mark Marissen; the allegations involved political corruption and selling influence. What those four names all had in common is that they all belonged to federal Liberals who were quite active, and highly effective, in advancing the interests of the federal Liberals as volunteers in 2003, and for many years before and after that time.

As their names were reported, it was also published that they were connected to Paul Martin and his leadership campaign. At the time the stories came out, I had not met Marissen and had not seen Bruce Clark for years. I knew Jamie Elmhirst, however, and did not believe for one minute that the young, idealistic true believer could possibly be associated with the allegations being made. As for Erik Bornmann, I had heard he was well respected in the party.

Mark Marissen provides a good insight into what happened on a human level when the BC Rail story touched his young family. "It was a morning, just after Christmas, and my mom was heading back to Ontario. I was about to give her a ride to the airport. Hamish opened the door to the cops. He was two and a half."

That was when Mark's plans for the day changed dramatically. He was shocked to find that the police were there to talk to him about BC Rail. Mark describes himself as having been raised, in his Dutch Christian tradition, to be innately respectful of authority in these kinds of situations, and he was very nervous. He immediately prepared to do exactly what he was asked.

Christy was not so agreeable, and her ferocious side came out. "Christy was furious!" he says, remembering that the angrier she got, the more he tried to keep the peace. "I was really nice to them, but she was right in their face. She asked them if they had already asked BCTV to show up."

This question was a reference to one of the most dramatic scenes in BC political history, where BCTV was on the scene when the RCMP showed up at Premier Glen Clark's house to question him about documents, and the BC public was able to watch the story unfold on the evening news.

The scene itself was an interesting parallel because the police had showed up to question Mark about documents, and later news stories reported there was a warrant, with allegations of criminal activity. "The cops knew she was deputy premier. They were coming for me and showed up with a letter saying that I was not under investigation. They had no warrant. They wanted some information about some folks."

The information they wanted was political, and the folks they were interested in were David Basi and Bob Virk, both of whom were staff with the Campbell government and had also been active in the federal Liberal leadership campaign for Paul Martin, which had just concluded, with Martin winning the leadership and becoming prime minister of Canada. Erik Bornmann was also active in the Martin campaign. The police wanted documents or emails that may have connected them all on an allegation of selling influence.

The police told Marissen that he was not allowed to share the letter they had given him, which said he wasn't implicated in their investigation, with anyone other than his lawyer. He asked, "Can I share with the prime minister's chief of staff? He's a lawyer."

Marissen convinced them it was necessary, and they said okay immediately. "At this time I was the volunteer campaign chair for the federal Liberals in BC. Although we weren't yet in an election, Martin had just become leader. So I agreed to co-operate if I could share the letter to save my job."

The allegations the police were investigating were that Erik Bornmann, a consultant, was getting information for clients through Basi and Virk and offering them jobs in Ottawa in return. "The cops wanted to know how this job recruitment process worked."

Marissen took time to explain the federal recruitment process in order to prove the allegations were false. Marissen knew that Bornmann would not promise jobs, and did not have the authority to give anyone a job anyway. As a campaign worker, he *could* legitimately submit names for consideration because it is actually a necessary standard practice after a successful leadership campaign; new leaders build their own teams out of the best campaign volunteers. Key workers like Marissen and Bornmann were *supposed* to send in names. But submitting a name didn't guarantee anything.

"If their profile showed they had potential, then there was a rigorous RCMP security check before anyone was hired. The cops said they didn't know that. That's when I started to get really nervous. Did these RCMP officers have any idea what they were doing?"[2]

Marissen searched his computer for the emails he had sent about the job application process, found them, and then made a point to state the qualifications of the people whose names were submitted. He gave the police what they requested. They had a few other questions, and then they gave him a copy of a draft press release, for reasons that were unclear to Mark.

"Perhaps they felt bad, and they just wanted us to be able to prepare for what was coming next. Scanning the headline quickly, I said, 'I think you've given me the wrong one. This press release is about drugs and organized crime.'

"The police said, 'No, it's the right one. Read further.'"

"That's when I knew that the shit was really going to hit the fan," said Mark.

That was day one of almost ten years of controversy.

The next few weeks were a very intense time, and the uneasy alliance of the BC Liberals with federal Liberals in BC came into play immediately in terms of who would take the fall for the controversy. Marissen laughs as he remembers just how starkly the lines were drawn between his role with Martin, and his wife's role with Campbell.

"We were heading to vacation in Mexico the day after the cops arrived, and in the cab on the way to the airport, Martyn Brown [Premier Gordon Campbell's chief of staff] was on the phone with Christy, trying to frame the BC Rail issue as a federal Liberal issue, and Scott Reid, Martin's communications director [from the Prime Minister's Office], was on the phone with me trying to make this a BC Liberal issue."

When the BC Rail story broke in 2003, it quickly found Marissen and Clark, even thousands of kilometres away.

"While on holiday in Mexico, I had to deal with the issue every day, including reviewing all the clippings that were faxed to us from Christy's office. It is pretty hard to have to wade through full media coverage of yourself, painting you as a criminal. I had to spend a couple of hours every day in the business centre responding to media, while Hamish and Christy were trying to have fun at the beach."

This story ran in the *Vancouver Sun* on December 30, 2003. Entitled "Marissen a top Martin organizer," its author, Petti Fong, writes about the RCMP investigation:

Mark Marissen says he's an innocent bystander in the 20-month-long RCMP investigation that led Sunday to a raid at the legislature and the retrieval of documents from his home.

As part of an RCMP investigation involving the politically appointed staff of two senior cabinet ministers, police asked Marissen Sunday for documents and correspondence.

It was a dramatic convergence of provincial and federal Liberal politics at the Port Moody home of Marissen and B.C.'s Deputy Premier Christy Clark, who have taken pains in the past to make sure their political paths do not cross.

The couple, who have a young son, are ardent federal Liberals who met in 1989 at Simon Fraser University. While Clark's path later veered into provincial politics, Marissen emerged as one of Paul Martin's top organizers and became his campaign chairman in B.C., in his successful bid for the federal Liberal leadership.

Marissen, raised in St. Thomas, a city south of London, Ontario, is now so eagerly sought by those who want to work in, be elected to, or simply do business with Ottawa, that he's thrown his cell-phone away because of too many calls.[3]

Marissen was very intent on correcting the public record, including reports that said the cops had seized documents from his home. There was an article in the *Vancouver Sun* that they would not correct, even after he called the publisher.

"I had to write a long letter including quotes from their editor in a speech he made at a conference about media losing credibility when they won't admit their mistakes . . . I think I got at least two dozen corrections or clarifications because of misreporting over the years."[4]

Of course, the corrections were never as high profile as the original story. Marissen wanted the media to retract anything and apologize if they said anything about search warrants, criminal investigations, or seized documents.

"We even managed to have Peter Mansbridge correct himself on *The National*."

After he corrected the public record as much as possible, the furor seemed to calm down, with the focus shifting to Basi and Virk.

A review of all the news stories over the years shows that Martyn Brown won the communications battle, for all the names in the media were directly connected to Prime Minister Martin, with Campbell's office barely mentioned. The overwhelming blame landed squarely on Christy Clark's shoulders because of her marriage to Marissen, and his questioning by the police. Martyn Brown would not be called to comment on this story until 2010 during the Basi–Virk trial. At that time, he denied any conflict of interest in Premier Campbell's office, and before the testimony in the trial could get into too much detail that might suggest otherwise, the case was settled; the Campbell government claimed the daily cost was $15,000 and it was not in the public interest to continue.

IN 2010, MINISTERIAL aides David Basi and Bob Virk pled guilty in a court of law after six years of maintaining their innocence. Two high-profile members of BC's Indo-Canadian community, their names were in the news continually in a very unflattering light.

Five years earlier, in 2005, I was in Vancouver to do an interview on CBC Newsworld about sexism and Belinda Stronach, who was a Conservative MP at the time, when I ended up in the elevator with David Basi and Bob Virk. I remembered Basi from his time in the legislature because he was known as one of the best political staff members. However, because he was with the Campbell government, I had no connection to him.

Basi recognized me and asked me to please understand that the news stories did not contain the true story of what happened, and that this would come out in court. I asked them how their case was going, and he told me that it was bankrupting them and their families. He said that they were receiving support from their extended families and that everyone had to mortgage their houses to help pay the legal costs, which were crushing them all.

The trial meant there was no way for Basi to earn money, so he was depending on the funds from cousins, aunts, and uncles, but their families wanted them to have justice. He said, eventually, when the full story came out in trial, people would understand who was at fault. He appeared very sincere in his comments to me that he was innocent.

Years later, when Basi and Virk entered their guilty plea, the government paid their legal fees, with many of the details not disclosed. I remembered our elevator conversation and thought that, although now the court testimony would not come out, the payment of the legal fees likely saved a lot of their families' homes.

I met Mark Marissen for the first time, very briefly, in the fall of 2007, when I was on the riding association for the federal Liberals in Powell River, and he was Liberal Leader Stephane Dion's national campaign manager, based out of Vancouver. I admit that the only association I had with his name at that time was that he was considered "the mastermind" of the BC Rail sale and that he was Christy Clark's husband. I associated him with Gordon Campbell and assumed he was a staunch supporter of the Campbell government, at the time in their second term. I expected him to be arrogant and disinterested in the idealistic side of politics. I was completely wrong. He was accessible, intelligent, and very much a federal Liberal.

In the fall of 2012, my husband and I attended a fundraising reception for MP Joyce Murray at the Sheraton Wall Centre Hotel in Vancouver. This is where I saw Bruce Clark for the first time in years. I had not seen him since his name had been in the news so prominently over the BC Rail controversy, and BC Rail was back in the news with a new investigation. I had heard previously that Bruce was not attending many public functions and noticed that he took a place in the background of the reception. I also felt people were avoiding him. I wondered if there was a stigma attached to him because of the BC Rail story, for that was the only time I had ever seen his name in the news, even after decades of political volunteer work. I had a good conversation with him and was reminded of his keen wit and great political instincts.

There was one issue that arose years after the eruption of the BC Rail controversy, and this was connected to Mark Marissen as a consultant. Marissen said that he found out one of his clients, CIBC World Markets, was overseeing the BC Rail bid assessments. CIBC World Markets is a big company, and there were no common points of contact between Marissen's work and CIBC's file connected to BC Rail.

However, Marissen says, "I insisted to Christy that she should declare it as a potential conflict, so that no one could ever accuse her of any possible conflict. This blew up in our faces, even though it was actually meant as the opposite."

After years of negative public headlines and ongoing stories of corruption linked to Christy Clark, Fraser Valley MLA John van Dongen filed a conflict complaint against her, initiating an investigation into these new conflict allegations. Marissen's consulting company had such a wide network that, in 2012, in response to van Dongen's complaint, Conflict Commissioner Paul Fraser himself had to declare a conflict, because his son had been a business partner of Mark's at one time. Fraser had to go out of province, and hired Gerald Gerrand, the former Conflict of Interest Commissioner from Saskatchewan, to conduct an independent investigation into the sale of BC Rail.

"Gerald Gerrand had to fly in to review the file and was astonished at all of the fuss," Marissen says. "Ironically, the van Dongen claim allowed for closure because it gave us a chance for a detailed response."

The report came out in April of 2013, just before the general election, completely clearing Christy Clark of any apparent or real conflict of interest.

The response from the New Democrats, who were leading in the polls, remained skeptical. NDP Justice Critic Leonard Krog repeated their party's promise to hold a public inquiry into the sale of BC Rail if they won the election. In 2013, most observers expected that the NDP would win the election. Krog expressed the NDP's intent to answer the question of whether the sale of BC Rail was corrupt.[5]

As it turned out, any lingering issues from BC Rail landed right back on Premier Clark's doorstep after the 2013 election. Even though she and her family had been cleared, the public outcry over the payment of legal fees for Basi and Virk remained and left a lingering impression of a payoff, since the fees were over $6 million. Acting Auditor General Russ Jones, appointed late in May 2013, undertook an audit of the fees and their process.

"In response to public concerns about the appropriateness of government funding the cost of the criminal defence for Mr. Basi and Mr. Virk," said Jones, "our Office initiated an audit of all special indemnities granted to individuals under the authority of the Financial Administration Act between 1996 and 2011."[6]

The acting Auditor General examined the twenty-six agreements that were available in this time period and noted that information was blocked due to solicitor-client privilege, which was unfortunate. Also, of the $11 million spent on over one hundred individuals' fees in this time period, $6.4 million was on the Basi–Virk case. Keep in mind that is only what the government paid for the defence; it does not include the cost to the government for prosecution. The cost of prosecution in terms of legal fees alone is usually higher than the cost of defence.

Jones's report concludes there was no political interference or inappropriate decision-making in the payment of the Basi–Virk fees and that the payment was consistent with the practices adopted for other government workers who were covered by this indemnity, which provided coverage of legal fees. The report states:

> In other words, the substantial amount paid to defend Mr. Basi and Mr. Virk was a reflection of the length and complexity of the legal proceedings rather than of government's administration of the indemnity agreements. The private legal costs paid to honour the indemnity agreements with Mr. Basi and Mr. Virk were significantly higher than the legal costs funded under any other special indemnity agreement. However, the other special indemnities we examined were not for criminal trials.[7]

Harvey Oberfeld, a retired TV reporter who spent thirty-eight years providing award-winning stories about politics and current events in Canada and BC, perhaps provides the best way to end this discussion on BC Rail. In his blog piece from January 2014 entitled "Full BC Rail Story NOT Yet Told," he offers his response to Russ Jones's findings.

> It's one of the oldest political, bureaucratic and corporate tricks: when there is bad news to release, do it late on a Friday afternoon or just before a major holiday.
>
> News has a shelf-life shorter than that of a day-old muffin. So what better tactic than to release embarrassing reports, documents than when so many are elsewhere occupied.
>
> Maybe it was just a coincidence that, on Dec 18 … just before the Christmas and New year break … BC's Auditor General Russ Jones released his report into the payout of $6 million to David Basi and Bobby Virk's lawyers in the BC Rail scandal.
>
> But I was not surprised at the timing: nor was I surprised at the findings.
>
> That still doesn't end the real story.
>
> The question in my mind, and I suspect a million British Columbians as well, is not whether the bureaucrats acted legally … it's:

What really happened to lead to the decision to accept a plea deal, such light sentences and have the taxpayers pay all of Basi and Virk's legal bills?????

What was the point of the whole $18 million trial/prosecution charade if it was to end so easily in a fix????

It stinks.[8]

After reading everything I could find on the BC Rail scandal, I feel like I witnessed a magician's trick of headline redirects. I suspect that the facts of the real story are now securely bottled up by solicitor–client privilege and tucked away in the various legal agreements associated with the plea deal. The premier at the time of the deal was not Christy Clark, it was Gordon Campbell. I feel comfortable stating that if the real story had been told, three names would be missing from it: Christy Clark, Bruce Clark, and Mark Marissen.

TOP Christy Clark at her campaign booth, running for office in the mid-1990s.
COURTESY OF CHRISTY CLARK

BOTTOM, LEFT Christy with newborn Hamish in hospital, August 2001.
COURTESY OF CHRISTY CLARK

BOTTOM, RIGHT Christy and Hamish at the beach. COURTESY OF CHRISTY CLARK

TOP Campaigning in 100 Mile House in the early 2000s. COURTESY OF CHRISTY CLARK

BOTTOM Christy Clark as the new premier, being interviewed by CBC's George Stroumboulopoulos in April 2011. COURTESY OF THE GOVERNMENT OF BRITISH COLUMBIA

TOP, LEFT Enjoying a moment with Hamish on holiday. COURTESY OF CHRISTY CLARK

TOP, RIGHT On a hike in the Gulf Islands with friend Dawn Clarke in the summer of 2010. COURTESY OF DAWN CLARKE

BOTTOM Cutting a BC Day cake with Premier Gordon Campbell in the early 2000S. COURTESY OF CHRISTY CLARK

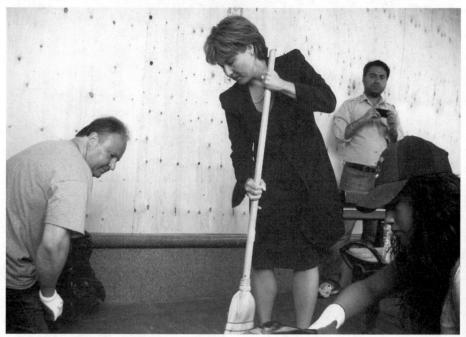

OPPOSITE, TOP Announcing a 30-unit housing program for single mothers in 2011. COURTESY OF THE GOVERNMENT OF BRITISH COLUMBIA

OPPOSITE, BOTTOM Helping with the clean-up after the Stanley Cup riots on June 16, 2011, the same day as the Board of Trade speech. COURTESY OF THE GOVERNMENT OF BRITISH COLUMBIA

TOP In Vancouver speaking at the Board of Trade as part of the Distinguished Speaker Series. COURTESY OF THE GOVERNMENT OF BRITISH COLUMBIA

BOTTOM In the premier's office in Vancouver, meeting with Her Excellency Miriam Ziv, Israel's Ambassador to Canada. COURTESY OF THE GOVERNMENT OF BRITISH COLUMBIA

TOP As host of the Council of the Federation of Canadian Premiers and First Ministers in 2011, Christy Clark enjoys a game of golf, laughing at Saskatchewan premier Brad Wall's moves. COURTESY OF THE GOVERNMENT OF BRITISH COLUMBIA

BOTTOM Governor General David Johnston and his wife, Sharon Johnston, visit Victoria in September 2011 and are presented with flowers by six-year-old Daisy Irwin, accompanied by Premier Christy Clark. COURTESY OF THE GOVERNMENT OF BRITISH COLUMBIA

TOP Premier Clark shows support for the BC Lions by doing a roar with the mascot and some supporters. COURTESY OF THE GOVERNMENT OF BRITISH COLUMBIA

BOTTOM Premier Clark pays respect to the Golden Temple in Amritsar, in India's Punjab state in November 2011. COURTESY OF THE GOVERNMENT OF BRITISH COLUMBIA.

TOP At the Williams Lake Stampede in 2012, riding with parade marshal Carey Price, NHL goalie for the Montreal Canadiens, who is from Williams Lake. COURTESY OF THE GOVERNMENT OF BRITISH COLUMBIA

BOTTOM Touring Hong Kong Harbour with the marine department manager, plus Port Metro Vancouver vice-president Duncan Wilson and Canadian Consul General to Hong Kong Ian Burchett. COURTESY OF THE GOVERN-MENT OF BRITISH COLUMBIA

TOP Premier Clark meets students at Northern Lights College in Dawson Creek in November 2011, announcing $794,000 to upgrade skills training equipment, as part of the provincial plan to align skills training with the job market. COURTESY OF THE GOVERNMENT OF BRITISH COLUMBIA

BOTTOM In Fort St. John in November 2012, Premier Clark is re-united with the first baby born there after she opened the hospital in July. COURTESY OF THE GOVERNMENT OF BRITISH COLUMBIA

TOP Premier Clark institutes a new statutory holiday for British Columbia in February 2011, and spends BC's First Family Day with her son, Hamish, at a special event featuring activities for families and kids of all ages. COURTESY OF THE GOVERNMENT OF BRITISH COLUMBIA

BOTTOM Premier Clark congratulates Tla'amin Chief (now Hegus) Clint Williams on reaching Final Agreement on their treaty, the fourth modern-day treaty in British Columbia. COURTESY OF THE GOVERNMENT OF BRITISH COLUMBIA

TOP Premier Clark and Deputy Premier Rich Coleman join young BC Lions fans to celebrate the 2013 Grey Cup Championship win. COURTESY OF THE GOVERNMENT OF BRITISH COLUMBIA

BOTTOM Bollywood dancing kicks off the BC India Globe Business Forum, preceding the Times of India Awards, just before the 2013 election. COURTESY OF THE GOVERNMENT OF BRITISH COLUMBIA

TOP Premier Clark visits one of the 45 schools benefiting from the announcement of over $584 million in funding to seismically upgrade high-risk buildings. COURTESY OF THE GOVERNMENT OF BRITISH COLUMBIA

BOTTOM *Vancouver Sun* columnist Vaughn Palmer attends a press gallery briefing at the legislature. Palmer is the longest-serving reporter in the gallery, starting his position in 1984. COURTESY OF THE GOVERNMENT OF BRITISH COLUMBIA

Love is submission, not decision ... The more you love,
the more you can love ... Love expands with use.
DAVID BROOKS

CHAPTER 14

HAMISH

ON AUGUST 25, 2001, Christy Clark's life changed, and she made
history when, about three months after the general election, she
and Mark welcomed Hamish Marissen-Clark to the planet, and
she blazed a path for other women seeking a career in politics. The event
was reported on the Canadian Press news wire:

> British Columbia's deputy premier has delivered a healthy baby boy,
> making her the first cabinet minister in the province's history to give
> birth while in office.
>
> Hamish Michael Marissen-Clark weighed in at eight pounds,
> six ounces.
>
> Clark plans to breastfeed her newborn while at the same time
> going head-to-head with opponents of her ambitious plans for the
> education ministry.
>
> The minister and another new mother working for the govern-
> ment plan to hire a nanny to care for their two children in a room near
> Clark's office.[1]

I cheered her on, one mum to another, because in 1992, when I was
twenty-six, I had been the first MLA to give birth while in office, and that

broke a few stereotypes. Clark was a re-elected MLA, a cabinet minister, and deputy premier, and she had navigated the 2001 election while pregnant and made it look easy.

"Being a parent teaches you patience more than anything else in the world," says Clark. "I remember reading an interview with Madonna, who said that becoming a mother was the most humbling experience of her life because she realized that she wasn't in control of her life anymore."

Clark was suddenly in love with the new centre of her universe, Hamish. Her perspective changed as she had to shift everything in her life and adjust her schedule around her baby.

Clark says that becoming a mother completely transformed her. "It made me more patient, it has taught me the meaning of humility, gratitude, and I have had more truly, profoundly joyful, transcendent moments in my life with Hamish than with anyone else. Children help you discover an undiscovered geography of your soul that you didn't even know was there."

Dawn Clarke enjoys watching Christy Clark's parenting. "When I see her with her son, Hamish, it is very reminiscent of what I saw with her parents, Mavis and Jim. She also brings her playful and funny personality into her role as mother."

Marissen says they planned a family. "It just took a while. We very intentionally wanted to have [a baby]. We would have had more if we were able to do it earlier. We had no idea of what we were getting ourselves into by having a child when she was deputy premier. We were both travelling so much, her mother wasn't well ... two kids would probably have killed us."

Once Hamish joined their family, they had to add many more logistics to their planning. They planned around her schedule, and Marissen would ensure his work could accommodate the legislature. "She would go to Victoria with him on Monday," says Marissen, "be back Thursday, and she would do night sittings. I had to be there when she was there."

Marissen worked from home while he watched his wife and child leave for the legislature. She was breastfeeding, and hired a nanny at her own expense. She shared the cost with another worker who had a child, and they brought a playpen for the children. "Meeting with Hamish" was on her schedule for breastfeeding.

"The NDP went crazy, and called it a taxpayer-funded nursery. It was ridiculous."

Christy Clark's personality also shifted as she adjusted to motherhood. "She became less of the 'rough around the edges fun partier,'" says Marissen. "She didn't seem very maternal before, but she became very maternal."

As for Hamish's early days, Marissen laughs: "There were so many soothers and various items found on the floor of Helijets and floatplanes... he's had a very unusual life."[2]

After handling the education portfolio and juggling her personal and professional life, Clark was shuffled to the Ministry of Children and Family Services. What did Vaughn Palmer think was the most memorable act from her time in Campbell's cabinet?

"Quitting. She said it was family. At the time it seemed more to do with Gordon Campbell saddling her with a no-win ministry. But that is what premiers do—hand out tough jobs to folks who should be able to handle them. Having said that, I can't think of much to say that would be positive about the Clark–Campbell relationship. Don't think they liked each other."

In her final speech as MLA, Christy Clark took time to reflect on the meaning of public service and interaction. Her remarks were made during her Throne Speech debate on Monday, March 7, 2005, and it is clear that she thinks it is her last time addressing the legislature:

> I have such a profound respect for the work that this Legislature does. I
> have a deep, deep love of politics. I love question period. I love debate. I
> love the people I've met. I even love the protestors. I love the conviction
> that every single member of this Legislature brings to the debate...

I was raised in a family that held public service in the highest esteem. My father was a teacher... To him it was the highest calling for any citizen to be able to serve in public office... He believed, as I do, that politicians have it in their power to make a better society if we make good decisions... He also knew that policy-makers have a huge role to play in building that great society.

We can give hope to families and help them put food on the table, pay their mortgages, hold their marriages together and give them time to nurture and love their children because they're not worried about their finances every day.

We can create a caring society where the least advantaged benefit just as much from our economy as those who are advantaged.

We can make choices today that will ensure our children have the same opportunities to succeed that we have... We can care for the elderly in our society and ensure they have dignity to the very end of their lives... We can give our children all of the knowledge they will need to compete anywhere in the world with the best and the brightest, on an even footing.

We can protect the rights of minorities from the tyranny of the majority, something that we must always be vigilant about in any society, if we want our children to inherit the rich democratic institutions we're working so hard to create. We can create a society where individual differences are not just tolerated but they're embraced and valued and recognized for what they are: the fabric of one of the richest societies anywhere on earth...

Every child in every community needs to be all of our children. That's how we will build strong communities.

Those were the values that my father instilled in me... That was what defined him as a B.C. Liberal through all of those years.

He ran as a candidate in three elections when he knew absolutely for certain that he would be defeated. He ran because he believed it was important to represent an alternative to British Columbia... It wasn't easy to be a Liberal in those days....

My father, when he was alive, revelled in politics. He watched question period every day. He pored over the newspapers. He sent me clippings. When I was living in France, he used to mail me these huge packets of clippings—mostly Vaughn Palmer's columns, so I can't say I was always well informed.

Unfortunately, my father died after I was nominated but before I got elected, so the task of helping me through my political life fell to my family. In my case, that was my mother, my brothers and my husband...

So I ran, and I was elected to represent Port Moody–Burnaby Mountain, and I learned the ropes in Opposition. I think if you look back at the Hansard record on those days, you'll see that I struggled to make a speech, and I wasn't very quick on my feet. But I learned through this chamber, as I've seen so many members do, how to be a better politician.

It was when we got into government that it really tested the bonds of my family. It was my husband, Mark, who made it possible for me to do this job. When I was invited to join cabinet and we had our son Hamish, it was Mark who travelled back and forth to Victoria every week—interrupted his work and spent three days here so that we could have dinner together every single night. It was Mark who emptied the rotten food out of the fridge and went shopping every night to make sure we had dinner to eat. It was Mark who allowed me to stand in the limelight while he stood in the back and didn't take any credit for the incredible support that he's been.

It was difficult to have a baby in the Legislature. I was shocked at the amount of controversy it created. I have brought my son here. I put him in office space that had already been allocated for me, and I paid for his caregiver. It was the public sector unions, who say they support more women getting into politics, that were protesting outside my constituency office, saying I was ripping off taxpayers.

My great-grandfather came to this province from Scotland. He left his family, his language, his farm—everything he knew—to come to British Columbia, because this was the land of hope and opportunity...

Mr. Speaker, I see that my time is up, so I want to close by saying this. I want to say that I have so tremendously enjoyed the opportunity to be a representative of my constituency...

I want to say that I will miss this place. I will miss politics. I will miss the hundreds of friends that I have worked shoulder to shoulder with. I've loved every minute of it. Other than being a mother, it has been the greatest privilege of my life to serve here. I hope that the MLAs who occupy this seat after me love this place just even half as much as I have.[3]

As surprising as her exit from politics was, some observers thought it was not a huge surprise that she re-engaged not long after her decision to not run again.

"Given her affinity for politics," says Moe Sihota, "and the extent to which it runs through her veins, it was surprising that she left politics to go on to the radio. Her statement that she did it to spend more time with her son is called into question by the fact that shortly after, she decided to run for mayor of Vancouver, and then leader of the party. It is understandable and noble if a woman says she is leaving politics to spend more time with a young child, but then it is quite puzzling that not long after, she is re-engaging. My guess is that she genuinely left to spend more time with her son, and shortly after, she missed the limelight."

At the time of Clark's departure, Marissen was working full-time on federal Liberal politics. "In 2006 we had gained in BC. We had the second highest number of Liberal MPs in BC history. There was a leadership campaign. I wasn't sure who I was going to support."

Marissen jumped into the fray to try to determine which candidate he would support for leader of the Liberal Party. "We hosted house parties to check out candidates. After the Dion party, I decided to support him. About three years of my life was taken up with that, where I was almost never home. Given the insanity of cross-country travel, it's no wonder that the Liberal Party hadn't had a British Columbian as campaign chair since 1968."

Marissen invested a lot of his life in his dedication to the federal party and was not involved in politics much provincially, except in Christy Clark's riding. Says Marissen, "She didn't get involved in politics federally, but she always voted Liberal. She let her federal membership card lapse in deference to the BC Liberal coalition. I didn't have to. I am my own person."

When Clark's mother was losing her battle with breast cancer in 2006, Christy dedicated herself to her mother's care. A friend of Christy's says, "When the time came, Christy did everything for her, and more. Christy gave her everything she needed. It consumed Christy's life for a while, but she never complained."

IN THE YEARS between Clark's exit from politics in 2006 to her re-entry in 2011, she was best known for her very popular radio show on CKNW that ran from August 2007 to December 2010 and her run for the Non-Partisan Association nomination for mayor of Vancouver in 2008, which she lost to Sam Sullivan.

Christy Clark and Mark Marissen were based out of Port Moody when she made the decision to run for mayor of Vancouver. Marissen still feels badly about Christy Clark's run for the mayor's seat.

"I didn't have an official position on this campaign, but I worked on it 24/7. I feel badly because I probably pushed her into it more than she wanted to run. Because I'm not from Vancouver, I didn't fully appreciate how people see Port Moody as a different place. I thought I lived in Vancouver. I also thought that if we lived near City Hall, we wouldn't have the logistical nightmare of Ottawa and Victoria in the mix of her life."

At the time that she decided to run, she didn't have a full-time job in radio. She had gone from a cabinet minister's salary with expenses to piece work and was earning about $300 a week. "She was getting ground down by having to manage the logistics around Hamish while juggling small jobs and not making any money."

In 2005, shortly after Clark appeared on the cover of BC Business, Vancouver-based political operative Marty Zlotnick approached her about running for the mayor's seat.

"We thought [Zlotnick] was the boss of the NPA when he approached her," says Marissen, "and expected that this meant that the party would do everything it could to support her in the nomination. I thought it would be easy. We will sign up way more people than Sam Sullivan regardless, and we will win."

After she announced, there was quite a reaction, and a lot of it was not positive. "In hindsight, it's good that she didn't win," says Marissen. "We were too cocky, and the structure of the vote favoured Sullivan. She lost by about sixty votes."

What was the long-term impact of her decision to run for mayor? Marissen thinks it gave her a lot of profile and "allowed her to be seen in her own right" outside of Gordon Campbell's influence.

Vaughn Palmer believes that the experience helped her later. "She would appear to have learned a thing or two about putting together a winning organization before her next entry into the arena."

When she began a full-time current-events talk show at CKNW, the largest news-radio station in the Lower Mainland, her media career took off.

"After she left politics," says Keith Baldrey of Global BC TV, "... I recall suggesting to both her and the station management at the time that she would be a good fit in that job. She was opinionated, well versed on a number of issues, and had a gift of gab. I was a regular guest on her segment, and in fact, I was often called upon if suddenly a scheduled guest fell through the cracks and didn't show. So we got to know each other fairly well in that capacity."

Moe Sihota thought Clark did well on radio. "I did listen to her, and I had a similar experience. When I left politics in 2001, I went on television in Victoria until 2004, so when I listened to her, I listened from that context. I didn't expect her to be non-partisan; I expected her to be fair-minded and partisan. You can't rinse your political label, so anybody that was on her show knew her political background, and anyone would have expected a bias. I think generally speaking she was accommodating of the left."

People in the political arena were more in tune with the program, and not everyone was quite so enamoured with her show. Guy Monty listened sometimes. "When she left public office and became a radio personality, her persona on the radio was infuriating. I know that she was hired to create controversy, but listening to the show, well it was very hard to take."

"When I look at her time on radio," says Sihota, "in many ways it reflected that which I saw when she was in Opposition and that which I saw when she was in government, namely that she understood the game of politics and was analytical in that regard. But in terms of depth around public policy, that wasn't her strong suit, and it hasn't been since she has become leader of the party."

To put Sihota and Monty's comments in context, CKNW talk show hosts tend to be more focused on politics than policy and have generally been on the conservative side of the political spectrum, as exemplified by Rafe Mair, Bill Good, and *Province* reporter Mike Smyth.

I was a guest on *The Christy Clark Show* in the fall of 2007, talking about politics in BC and life after politics. I had not seen her for several years, and to be honest, was not sure how I felt about her because of her time in the Campbell government. As soon as I saw her, it was as if the years fell away, and there was the same Young Liberal I had known in the hungry years. She was just as passionate and smart during that interview as I remembered. When we were off air, I asked her about her little boy. She beamed about how he was the best thing that ever happened in her life. He was six years old then, and as is my way, I started nagging her about having another baby.

"I'd have to see my husband to conceive another child," she replied.

According to Christy, Mark was always travelling, and since that was in the Dion years, I'm sure his feet barely touched the ground. The following year, after the 2008 federal election, they decided to divorce.

Although it was not finalized until 2014, they remained friends throughout their separation and divorce.

"We grew apart," Marissen said. "We just hadn't spent time together. I found the nearest condo to the house, a five-minute drive away. Hamish is with me half the time and her half the time. In some ways it worked out well for Hamish. He has more undivided attention." In March 2016, Marissen married Maryam Atigh, a beautiful Persian-Canadian woman from North Vancouver, whom he met at a Liberal function.

Derek Raymaker remembers an encounter with Clark during her radio days. "In about 2009, she was out of politics, doing her radio show on CKNW, and I was in Vancouver for a couple of days. It was the first time I met up with her after her split from Mark. We met for coffee after her radio show. As we left the station, we came across three homeless men, [who] she knew by name. They were like her buddies, and she was their buddy. They had a strong friendly relationship that went beyond asking for change. She knew everything that was going on in their lives. One of them hugged her, and she hugged him back. She let them know about things that were out there that might help them out. Can you imagine Gordon Campbell or Adrian Dix doing that?"

According to Raymaker, "She goes out of her way."[4]

The radio show on CKNW was largely seen as a huge success. Live radio is a challenge, and a daily show means you need to have solid content. Long-time friend and political ally Mike McDonald remembers the radio show well.

"She was a natural on the radio. In the summer of 2010, she encouraged me to do a political panel with her. I recruited a right-winger, Jordan Bateman, and a left-winger, Lesli Boldt, and we started a provincial political panel in October 2010. Little did we know that almost immediately Gordon Campbell would be in serious jeopardy and ultimately resigned, triggering the leadership race. From one week to the next, we would be handicapping the contenders. Would it be Carole Taylor? Would it be Dianne Watts? The conversation didn't turn to Christy for a few weeks. When she decided to consider the opportunity, she took leave and my radio career ended!"

In the weird serendipity that is BC politics, the day that Gordon Campbell announced he was resigning as leader of the BC Liberal Party, Gordon Wilson was a guest on Christy Clark's radio program. While he was on hold, waiting to go on the air, he heard the news of Campbell's resignation through the phone line.

Said Wilson, "Before she could even ask me a question, when she said, 'Hello, Mr. Wilson,' I said, 'Christy, are you going to run?' She replied, 'Who is doing the interview, me or you?' and then, 'I have no plans to run.' I laughed at her response and said, on the air, 'Okay, then your answer is yes.' She asked if I thought she *should* run, and I said, 'Absolutely.'"

McDonald says that, while Clark was on the air, "She criticized her former colleagues, which put some noses out of joint. She consulted Jimmy Pattison [CEO and owner of the Jim Pattison Group] before she ran for leader. He told her, essentially, to connect with voters like she did on her radio show."

Says Baldrey, "When Campbell announced he was leaving and the job was opening up, Clark phoned around to a bunch of people asking them what they thought of her entering the race. I have to confess when she phoned me my reaction was basically, 'Are you nuts?'"

He knew from his years of watching BC politics that it was a volatile environment.

"I thought she had a great and influential job at CKNW and was making pretty good coin there. The political prospects of the BC Liberal government seemed shaky at best and doomed at worst, and I thought anyone becoming leader of the BC Liberals might be looking at short-term employment."

Thinking back to that conversation, Baldrey says it served to remind him of "what kind of political animal Clark really is. She lives and breathes politics and she once told me that once you get a taste of what it's like to be elected, there's really no comparison to anything else and you want to feel that experience again and again. I'm not a politician, so I don't know what that feels like, but when she finally made her decision, it was obvious the political animal in her ruled the day."

BRITISH COLUMBIANS HAVE had a chance to watch Hamish from the time he was born, in political vignettes, growing up before our eyes. He was born to a cabinet minister, attended school while his mom was a famous radio host, and was nine years old when he joined his parents to watch the results of the leadership race.

Being the son of the premier is not easy, and Hamish is quite high profile in his mother's career. This means that he is constantly exposed to people who have strong opinions about his mom.

By the 2013 election campaign, Hamish was eleven years old. In the 2013 West Kelowna by-election, as Hamish was knocking on doors with his father, Marissen remembers one challenging comment.

"A woman answered the door and told us she wouldn't vote for Christy Clark because she gave BC Rail to her husband."

Marissen thanked her, and they walked away. This comment led to a discussion between father and son about why they just walked away, instead of arguing with her.

"Hamish was confused why I didn't defend myself. I explained that with people like that, there's no point in causing an incident and it's just better to go on to the next door. He still thought that I should have engaged with her. Perhaps he was right. I was thinking like a political organizer and he was thinking like a human being."

In 2016, Hamish is an intelligent, engaging, and funny young man. At fourteen, he is tall and slim, with thick dark blond hair that falls forward over his face. He looks like he could walk on stage with any popular boy band and grab the microphone, and he is certainly handsome enough to draw the eyes of the young ladies.

Hamish's legal last name is Marissen-Clark. If people wonder about Hamish's last name, they can ask his father about it. Says Marissen, "I thought it sounded better, and it never occurred to me that people would assume that meant his last name was Clark."

The name Hamish is connected to men's names on both Christy and Mark's families' sides. It is an old Scottish name that means James, and so Hamish is named after Christy's father, Jim, and Mark's brother James.

James also originates from the Hebrew name Jacob, which was Mark's grandfather's name.

What about the impact of politics on her son? Dawn Clarke has a good perspective on Christy Clark, the mother, and how Hamish is handling the unique life he has had.

"It will be interesting to see what Hamish decides to do with his life; he appears to be a very well rounded individual already."

Recovering from suffering is not like recovering from a disease.
Many people don't come out healed; they come out different.

DAVID BROOKS

CHAPTER 15

RE-ENTERING
POLITICS

O N NOVEMBER 3, 2010, Gordon Campbell announced his plan
to resign the leadership of the BC Liberal Party and stay on as
leader until his replacement was elected by the membership. The
announcement shocked many political observers because it was about a
year and a half after the election.

At the time of Campbell's resignation, he was the longest-serving
Canadian premier and the first BC premier to serve more than one term
since Bill Bennett (1975–86). He'd been premier since 2001 and leader of
the BC Liberals since 1993, a seventeen-year stretch.

In the fall of 2010, Gordon Campbell's popularity was at a record
low of 9 percent in the polls, and there were rumblings of an imminent
revolt by his caucus members. One high-profile issue that brought down
his popularity was the introduction of the Harmonized Sales Tax, or HST,
which provoked a large anti-government populist uprising, led in part by
former premier Bill Vander Zalm.

When Campbell made his resignation announcement, the BC Liberals were behind the NDP in the polls. Carole James was leader of the
NDP and was facing her own party's internal rumblings. James had
been leader of the NDP since 2003, and although she had dramatically

improved the party's standing in the legislature, she had twice faced Gordon Campbell in general elections, and he had led his party to majority government both times.

While this turbulence was occurring in BC's political arena, Christy Clark was covering all of it, and more, on her radio show. Christy Clark says that the idea of running did not occur to her at first. Clark had come out of a lean financial time in the private sector and was making some real headway in her new career.

"I was building a career in radio and media. I was working really hard to learn the business, to get better at it. I was enjoying it, and I thought I could really build it into something," Clark laughs. "I get frustrated with these depictions of my personal history where they say that my whole life I was planning to run for premier. It's just not true."

When the leadership campaign began, Christy Clark was separated from Mark Marissen. Clark's brother Bruce Clark and Mark Marissen were having coffee together when they saw Campbell's resignation announcement on TV.

"We both looked at each other and said, 'Wow, Christy should run for the leadership!' Within an hour or two of the news story of Campbell's resignation, she was on television saying 'absolutely not' to the idea." Marissen laughs. "So we decided that she needed an intervention. Later that evening, Bruce went to see her, and she basically agreed to *think* about running and to stop saying that she wouldn't do it."

They were men on a mission to change her mind. Bruce and Mark divided up the first steps to try to convince her to run.

Says Marissen, "Bruce decided to find out if he could put people together and raise the money for the campaign. My job was to find out if she could get the support of the people in the party and in the public, and she promised to figure out what she wanted to do about her job. The idea of running was not on her mind at all; it took a few weeks for her to even take it seriously."

Clark confirms that she was not thinking about going back into politics as she covered the political stories on her radio show. "Originally, I didn't think it was something I would do because I thought there would be other candidates who could win the election who would step up."

Marissen's polling work was key to proving the need for her to run. He hired Paul Martin's pollster David Herle and tasked him with determining the public's perspective.

According to Marissen, "The poll showed that she was, by quite a margin, the most popular candidate with the public. We managed to get our hands on a rural, suburban, and urban membership list for the Liberal Party from a friend, so we polled those three as well."

Marissen says the poll results were confidential because Christy still had her job and hadn't decided to leave it. Marissen and Bruce Clark kept the pressure up. Bruce Clark's fundraising enquiries found there was a lot of appetite for her to run and that raising money for her campaign, as a result, would be straightforward.

Says Bruce, "I was very confident that she would change her mind, and run."[1]

"As the leadership campaign went on," says Christy Clark, "a lot of people in the business community and the BC Liberal Party came to me and said, we need a candidate who can win the election, and you need to run. There were people I really respected who kept saying that I had to run."

There was a concerted effort underway to draw her attention to the need to challenge the NDP. Clark's initial belief that a suitable, and *winnable*, candidate would step up didn't happen.

"I don't mean to disparage the other candidates who did run," she says. "They were all really good, but the most common comment that I heard was that the BC Liberal Party needed a refresh, and we could only do that if we had someone who was seen to be an outsider."

Christy Clark's radio profile made her a popular potential candidate who was also perceived as an outsider, especially in terms of big issues like the HST.

The pressure began to have an effect on her, and she started to consider the leadership campaign in the context of the next election and her view of the options for government at that time.

"When I was in politics, I had been so committed to making sure we defeated the NDP because I was focused on the economy and jobs. After devoting nine years of working hard for the province, I couldn't imagine sitting back and letting the work that I'd done fall away."

SHORTLY AFTER CAMPBELL announced his resignation, Liberal caucus members began to announce their intention to run.

The first was Moira Stillwell, MLA for Vancouver–Langara. Stillwell had also served in Campbell's cabinet. Her campaign did not receive any caucus support and ended about two weeks before the leadership vote. When she backed out of the race, she endorsed George Abbott.

George Abbott from the riding of Shuswap was the second candidate to announce. Abbott, originally elected in 1996, had served in Campbell's cabinet as minister of aboriginal affairs, of education, and other positions. He had strong caucus support from MLAs and announced on November 25, roughly three weeks after Campbell's resignation announcement.

On November 30, MLA Kevin Falcon announced his intent to run. Many viewed Kevin Falcon as the·frontrunner. Falcon represented the riding of Surrey–Cloverdale, had served Campbell as Minister of Health and Transportation, and had strong caucus support from some of the MLAs considered to be power players in government.

Mike de Jong, MLA for Abbotsford West, was the last MLA to declare his intent to run. He had been a strong player in the Campbell government in senior cabinet positions. He announced on December 1 and had the support of two caucus members.

The mayor of Parksville, Ed Mayne, entered the race late, on January 3, and withdrew on February 17, endorsing George Abbott.

On November 27, 2010, Christy Clark announced, on her radio show, that she would decide whether or not she would seek the leadership of the Liberal Party within a week. On December 8, she made it official.

Clark says her family life was on her mind as she was lobbied to run.

"I spent a lot of time worrying about my parenting commitment to Hamish. He was getting to an age where he was starting to get really busy with other activities, including sports and school, so that helped."

The lessons her father had taught her about public service also played a part in her consideration. "When I was trying to decide what to do, I realized I felt a strong sense of obligation or duty. The opposite of what I wanted Hamish to learn would have been that I turned away from doing something because it was too hard to do. I don't think that's the right example for your kids."

Some people wondered if her decision to leave politics in 2005 would haunt her, or if she delayed her decision because she was not sure how to present her change of mind. "The hardest part of the decision was the Hamish part, not explaining why I left and why I came back. I knew I could lose the leadership, and it seemed a certain likelihood, according to the polls, that I would lose the election. But I wasn't sure what the best decision was regarding Hamish."

Some friends tried to talk Clark out of running. "We had a lot of conversations about how to evaluate the decision to run," said Mike McDonald. "I knew that a lot of people were encouraging her to run. I believe she made that decision ultimately because she believed she was the best option for the party and did not want to shirk her responsibility. She has a strong sense of history—she is a devout British Columbian. The opportunity to serve and the duty to lead is flowing in her blood."

Mark Marissen finds it ironic that Mike McDonald, who was more in touch with the BC Liberal Party's internal politics at that time than either Mark Marissen or Bruce Clark, thought that Christy shouldn't run and that it would be very difficult for her to win.

Says Marissen, "I got the sense that he thought that we were kidding ourselves. I think he thought that Bruce and I were shit-disturbers. Mike is more of an establishment person. But he agreed to run her campaign because of his personal relationship to Christy."

After many discussions with people close to her, friends, family, and associates, she decided. "On balance," says Clark, "it was equally important to set an example that I would work as hard as I could to make a difference in the world. In the end, it was the lesson I wanted to leave [Hamish] with that convinced me.

"When I thought about it, I thought when you are confronted with a call to action to do something good for your province, you cannot turn away. I had the opportunity to take this job because I believed that my party and my province needed me."[2]

"When I entered the leadership," says Clark, "I was a lot more convinced I would win the leadership than win the election." This conviction was after weeks of talking to business leaders and party members.

"By the time I declared, the major candidates were all in the race. I had a huge reach into homes because of the radio, and it helped. When I travelled to places like the mid-Island, people would come out and sign a membership to the party who had never held one in their lives because they felt they knew me from the radio show."

Media observers thought Clark might be making a mistake. "I thought she was very good on radio and couldn't understand why she would want to give up a great gig for such a risky one," says Vaughn Palmer. "Overlooked how much she had the political bug. There was that great line from her about politics being like an old boyfriend."

Palmer is referring to a 2005 *BC Business* article where Clark says:

> "In the first six months after you leave you still remember all the reasons why you left," she says. "And then a couple years down the road you're sitting alone at night by yourself in your living room, maybe into a glass of wine, and you're thinking, *God, that guy was great! I miss him so much!* And you pick up the phone and dial.[3]

"That's it," says Palmer, "Clark's sense of humour and self-knowledge, all rolled into one."

Dawn Clarke thinks Christy Clark really had no choice when it came to running. "She's a person who has always followed her gut and has tremendous instincts. Initially, I was a little surprised when she decided to run for the leadership, but then I also wasn't. Although she was a charismatic and intelligent radio host, she is, more importantly, a leader at heart, and one who is thoughtful and likes to effect change. Also, timing in politics is everything, and she was ready, and this was clearly her time."

At the time of Clark's decision to run, Moe Sihota was in private business and was also the president of the BC NDP. "It was apparent to me that, because of her work on radio and her personality, she was looking for the opening, and it came and she leapt at it. From the outset, I thought she would win. I saw that most of her colleagues would not support her, but I always knew she had the organizational machine that was at her disposal."

Mark Marissen was a big part of setting up that machine at the beginning of the campaign. As devout a Liberal as Marissen was, he had learned that in BC you have to protect the free-enterprise coalition in order to have a chance of maintaining power.

"We were concerned that everyone would see this as if she was a federal Liberal candidate, so we needed some [federal] Conservative credentials," Marissen says. "We couldn't find anyone in BC who was a prominent Conservative who could help her. They were already with the other candidates, especially Kevin Falcon."

As a result, they had to shop outside the province.

"I had met a guy named Ken Boessenkool when I was in Ottawa, chairing Stephane Dion's campaign. I had faced off with him in a debate at a think tank conference right after the 2008 election. He was Harper's guy, and I was Dion's, and we argued about everything."

Marissen remembered that even though they argued a lot, Boessenkool and Marissen had a similar upbringing in the Dutch Calvinist community. Boessenkool remained a stalwart Calvinist and a fan of Christy Clark.

"He really liked that she was fiscally conservative, plus he supported her education policies." Ken's father was the head of the BC association for independent schools. "I disagreed with Ken on many, many, many things, but I knew we needed to bring in a high-profile Conservative

so that Conservatives within the party could feel comfortable supporting her."

The move to bring Boessenkool in was controversial. "The main criticism levelled by the Kevin Falcon supporters was that Christy was going
to destroy the coalition. It's ironic that Kevin's people were saying this,
given that Kevin was part of Christy's coalition a long time ago in student politics at SFU."

Boessenkool was keen to be involved. He was from Alberta and had
strong Harper-related Conservative credentials. "He was going to move
to BC and work full time on the leadership campaign, but there was so
much blowback from federal Conservatives in BC when they found out
his plans that he soon told us he couldn't help at all. He cited some kind of
policy that his role with Harper's team prevented his involvement."

Marissen laughs even as he talks of the frustration of having this
key player removed from the team. Fortunately, during the time that
Boessenkool was planning to be involved, he had introduced Marissen
to Dimitri Pantazopoulos, a federal Conservative from Ontario.

"Dimitri brought in others, including Stockwell Day and Preston
Manning," says Marissen. "Plus, in late December, when Rich Coleman
announced that he would not run for the leadership, but would instead
support Falcon, we picked up some other key local Conservatives like
Patrick Kinsella and Sharon White."

It was a campaign with strong competitors and huge stakes, for the
party was government and there was a two-year run to the next election. Some commentators stated that cabinet minister Kevin Falcon
was expected to be Campbell's successor, others that Christy Clark was a
born winner. Once she was in, she was focused on the win.

Says Moe Sihota, "She ran against Kevin Falcon and George Abbott,
who were both more inclined to policy than she was. They lost to
her because she has a sense of the game of politics; she's a gifted
communicator."

Kevin Falcon, elected in 2001, was an active member of the Social
Credit Party in the 1980s, and became involved in municipal politics in

Surrey, organizing for right-wing candidates including Dianne Watts, former mayor of Surrey and elected as a Conservative MP in 2015. Falcon was best known, prior to his election as MLA in 2001, for heading up a political campaign called "Total Recall," which was aimed at recalling the NDP MLAs. Initiated after Glen Clark's surprise win as premier in 1996, this was a highly negative campaign. Falcon was very popular with the right-wing of the BC Liberal Party.

George Abbott, elected to represent the riding of Shuswap in the Interior in 1996, was a political scientist and served on the Columbia–Shuswap Regional District before his election as MLA. I always think of George Abbott as "Shep" because of one late-night sitting in the 1990s when he delivered an ode to his dog Shep in the legislature. It was likely just filibustering (making long speeches to draw out debate), but he'll always be Shep to me.

HOW DID CLARK approach the campaign? She had a great team, and she is a natural in politics. The key, according to Clark is "making sure you sign up more members than the other campaigns, and getting the vote out, there's a method for that. It's not rocket science. I was really lucky. I was surrounded by really smart people."

Mark Marissen, Bruce Clark, and others who supported her in the past and helped with her success all showed up again, and new key team members were recruited.

Once Christy Clark made her decision to run, Marissen was told by Mike McDonald that he had to take a background role because some people thought he might alienate other supporters.

"What I thought was ironic about all of this is that Mike McDonald thought I was too liberal and would scare off all the Conservatives, but I was the only one who brought any Conservatives in, at the outset." McDonald was campaign manager, so Mark didn't kick up a fuss.

Derek Raymaker worked on Clark's campaign and says Clark's personality really resonated with the members of the party as they travelled the province seeking support. "She's very, very generous and authentic.

She can be overgenerous and over-trusting. If you have a conversation with her, you will talk more about yourself, because she cares. When she goes door-to-door, she is asking people what they think, and she is processing their answers so she can move down a path to help people."

Former Liberal leader Gordon Wilson was a bit more cynical. "I wasn't a member of the Liberal Party. To be quite frank, I thought the Campbell resignation timing showed that the fix was in and Kevin Falcon would be the successor."

The choice of leader was seen, by Wilson, as pivotal in determining the future of the party. "The leader of the party has a key role to play in directing party policy. It was a huge opportunity to take a clean break from the Campbell years. I knew Clark to be a Liberal. I hoped that she had not sat in the Campbell teapot so long that she was steeped in conservative thinking. She was an outsider by this time."

This "conservative teapot" may have played a part in shaping Clark's winning team as the new free-enterprise coalition by making room for Sharon White, who took a lead role in Christy Clark's campaign. "We knew we could win. I believed she was the one to win. I felt we were one of the frontrunners from the beginning."

Sharon White is a Queen's Counsel lawyer and a partner at a prominent Vancouver law firm, Richards Buell Sutton LLP. She is a powerhouse of a career woman in terms of her accomplishments. In addition to having a reputation as an excellent lawyer, she is chair of the Real Estate Development Group at RBS, past chair of the Securities Section of the BC Branch of the Canadian Bar Association, former member of the Securities Advisory Council of the BC Securities Commission, former director of Partnerships British Columbia, and former director of Farm Credit Canada. What this bio doesn't tell you is that she is a ladylike, soft-spoken person who is kind-hearted and loving toward animals and has a delicate, Audrey Hepburn-like elegance. She also has a keen sense of humour.

White is also a true believer Conservative with a very strong sense of the need for integrity in politics. Her network and organizational skills in the leadership campaign were critical. Says Sharon White, "I think because I come from the conservative side of the political spectrum, I provided a good balance and showed we could all come together under her leadership. When she ran for the leadership, I was asked if I would co-chair that campaign with Mike McDonald."

White said yes. And so the Liberal–Conservative coalition was properly represented in Clark's campaign.

McDonald remembers the leadership race with excitement. "I helped populate the campaign committee, and from December 13 to February 26, 2011, I was 24/7 to help elect Christy. We were a campaign of volunteers. We had one paid staffer, and the rest of us just did what we had to do. It was amazing the calibre of people who came off the street to help out, whether it was in advertising, communications, policy, or tour. The volunteers who manned the phones, inputted the data, and worked in the ridings were a first-rate team and they were there because of Christy's magnetism and the enduring relationships she had built across BC."

In the press gallery, Keith Baldrey of Global BC predicted a Clark win. "I didn't think anyone could match her in signing members up, and none of the other candidates could match her charisma and star appeal, which goes a tremendous way when you're trying to woo someone into joining you. She also had a pretty good campaign team behind her. When I started hearing stories about her going into a series of coffee meetings with thirty or forty people and cajoling and wowing all of them, I knew it was basically over. She also seemed extremely confident through the whole leadership campaign, and her rivals did not."

While the campaign was underway, the party agenda shifted, and Clark's leadership style emerged.

"I was heavily involved," says Sharon White, "and it was a fantastic experience. I look back on it and think that all of our lives have changed so dramatically in the five years following. That was the first time that I saw her as she is in 2015, as premier. There were glimmers of her leadership before, but it was during this campaign that her strengths emerged."

Christy Clark started leading her campaign as she would later lead the party and the government: she was inclusive. White says it was empowering to be able to be part of the strategy in meaningful discussions and know that what was decided would be implemented. Says White, "She had a voice in these [strategy] discussions, but once it was agreed, she would go off and put it into action. She was open to taking direction from her team, provided she was comfortable with the direction."

Coming back from the outside created some tensions between Clark, who had left after one term in government, and the MLAs who had stayed. She had almost no caucus support, and there were some tensions between her team and the other two leadership rivals, both of whom were sitting MLAs. Clark's vocal and high-profile radio talk-show host position had also meant that hers was one of the voices critical of the government that these MLAs represented after she left.

Says McDonald, "I think the main reason she didn't get caucus support is because that caucus had gone through a lot together, most recently the HST issue, and Christy was from outside. The fact that she had not been there over the past six years when they had gone through this tough time meant they wanted to support someone who had been through it with them."

Being an outsider had some advantages, however. Clark's decision to run caught the attention of some people who had previously been engaged in politics but had since walked away. Former Liberal Party of BC President Floyd Sully says when he heard of Clark's decision to run, "I was ecstatic. Wow, we have the best person in the province going for it. It was way past time that we had someone back in the leadership who was a true Liberal."

How do you re-engage with a party as an outsider, especially when you do not have a seat in the legislature as an MLA?

"She out-organized them," says Gordon Wilson. "She brought new people in, worked the ridings, went into areas that were not conventionally the Campbell areas, and made decisions that delivered the members.

She had a message that was different. Falcon's message was status quo, whereas she talked about a time for change and family values and issues that were relevant to the BC voters."

Speaking of what was going on behind the scene, McDonald says, "Our pollster, Dimitri Pantazopoulos, had polled the members in December 2010. We learned that Christy had a substantial lead among members. Grassroots members wanted change, especially in ridings where there was no BC Liberal MLA. We won thirty-five of thirty-seven ridings where there was no BC Liberal."

There were eighty-five ridings at that time, so just by focusing on these "non-Liberal" associations, Clark's team was close to victory.

Mark Marissen says the decision to focus on the ridings with the fewest party members was part of their strength because the leadership race had votes set up on a points-per-riding system, rather than one member, one vote.

"The best place to target are those ridings with the lowest membership because if a riding had few members, the point system was divided by fewer people. For example, if every riding had a hundred points, then adding supporters in ridings with almost no one meant our supporters had a good chance of capturing most or all of the points."

He put his efforts into a targeted base of members in the Indo-Canadian community. "The thing about the Indo-Canadian community," says Marissen, "is that it is extremely active politically and can be found everywhere, including ridings where there are very few members, so that was where I spent most of my focus."

Christy Clark may have been popular with the party members, but her lack of support from the elected MLAs created an unfortunate impression.

During the leadership race, there was strong competition for support, and some candidates tried to scare members into voting for a candidate that would be best at keeping the free-enterprise coalition together. Vaughn Palmer says the leadership race had a nasty tone. "Both Falcon and Abbott and their organizers played on that lack of support. The Falcon camp also pushed the line that Clark, being a Liberal, would alienate the Conservatives in the BC Liberal coalition."

Keith Baldrey did not think this BC Liberal leadership race was particularly vicious. "I've covered about a half-dozen leadership races at both the provincial and federal levels, and I've always thought internal party battles could often get more heated than actual elections. But I don't think the BC Liberal leadership race stood out for its nastiness, certainly not like some of the others I've seen."

He acknowledges, however, that it wasn't all smooth sailing. "There was some tension for sure, and the candidates occasionally took shots at Clark, but I don't think anything was particularly bitter or vengeful. I suppose Abbott and Falcon were the most critical, but even then a lot of it was couched in careful terms."

Says McDonald, "While it was disappointing that she did not garner more support from the caucus, we turned it into an advantage. In early January, I received a call from Mel Couvelier, who was a Liberal who had served as Social Credit finance minister and two-time leadership candidate. When I told him we had hoped to have more MLAs support us, Mel said, 'Don't waste your time. It's not about them. Focus on the people.' Mel's clarity on this issue helped remind me that Christy's appeal was indeed 'out there.'"

Mark Marissen remembers the night of the leadership vote. The convention centre was booked, and there was a stage where each ballot was announced. Voting by members was conducted via telephone and Internet, and there were side rooms for each of the leadership candidates. Marissen does not remember too many supporters in the main hall, and none on stage, where the caucus awaited the results.

"She didn't have as many people in that room as the others did, so it wasn't as excited a crowd when the final result was delivered. It was a bit tense because we weren't as far ahead on the first ballot as expected, although we did expect to win."

Clark recalls the night of the vote clearly. "I never allow myself to assume I'm going to win, ever. The leadership race went to three ballots. Hamish said to me beforehand, 'Are we going to win?' I said, 'I think we have a good chance, but if it goes to more than two ballots, it's not a good sign.'"

After the first ballot, Mike de Jong was dropped. After the second ballot, George Abbott was dropped. It was down to Christy Clark and Kevin Falcon, with Christy Clark in the lead.

Clark says that Hamish was following the ballots closely. "When it went to three ballots, he came and said, 'I feel really sick right now' because he knew what that meant. Then Falcon's ballots came in, and I won. It unfolded, as these things often do, quite differently than expected."

In the end, the final vote was Christy Clark with 52 percent and Kevin Falcon with 48 percent.

Once the votes were announced, it was time for the new leader—and new premier—to take the stage in the convention centre and make a victory speech.

Marissen remembers. "It was a bit uncomfortable for her to take the stage because there were not a lot of supporters on that stage, and the room was not very excited with the leadership result, as one might expect. The stage was full of the other candidates and the caucus, and she only had one MLA supporting her."

When asked if he was surprised at Clark's success over Falcon, Vaughn Palmer says, "A bit. Though she was leading in all of the internal polling on the strength of both the freshness of the campaign and her not having any connection to the HST debacle and other late Campbell follies.

"The Clark and Falcon teams were both impressive in organizing, particularly in the Indo-Canadian ridings, which was also a big factor in the NDP leadership at the same time. Abbott lacked an organization in that community... but by rallying MLAs who were on the outs with Campbell (who tacitly and not so tacitly wanted Falcon), Abbott split the anti-Clark votes and probably made the difference for Falcon. When

Abbott dropped from the ballot, Falcon didn't pick up the second choices he needed to defeat Clark."

When asked how she felt the moment she realized she had won, Clark says, "It wasn't unalloyed joy because I knew there were a lot of people out of the leadership who would not welcome me as their leader, and I knew many of them were in the caucus."

Christy Clark won the leadership of the BC Liberal Party on the evening of February 26, 2011.

She was the second female Liberal leader since the party's inception in BC in 1902. Shirley McLoughlin (1981–84) was the first. McLoughlin, a teacher by profession, was not elected to the legislature. Shirley McLoughlin had won the leadership when the party was in disarray, and she worked hard to help rebuild it. When she won, there were fewer than five hundred members voting.

Christy Clark won the Liberal leadership when the party was in government, with 52 percent of 8,500 votes cast. Christy Clark was not an MLA at the time that she became premier, and one of her first priorities was to run in Gordon Campbell's riding in a by-election.

Christy Clark was BC's second female premier, Social Credit Leader Rita Johnston being the first. Johnston won the leadership in April 1991 and held the position until November of that same year. Christy Clark would contest the next election in an attempt to become BC's first elected female premier.

We can shoot for something higher than happiness. We
have a chance to take advantage of everyday occasions to
build virtue in ourselves and be of service to the world.
DAVID BROOKS, *The Road to Character*

CHAPTER 16

ALONE IN
THE CROWD

O
N THE FIRST day of class, my second-year political science professor, Dr. Powers, asked a question of us, "What is politics?"
Over one hundred eager faces looked at him, trying to figure out
how to answer his question and make a good impression. There were a
few students who even took stabs at an answer. Finally, he put us out of
our misery.

"Essentially," he said, "politics is the answer to the question, 'Who
decides?'"

From there he went on to teach us about the basis of power, which is
the interplay, competition, and ultimately the ascension of the individual who is named as the person who decides.

Historically, in both Canada and in British Columbia, the answer to
"Who decides?" has been, "He does." Which is to say our governments at
every level were represented by, basically, white guys. Women fought
and succeeded to obtain the vote in BC in 1917 and federally in Canada
in 1918. The right to stand for office was granted in 1919, and eligibility
to serve in the Senate came in 1929. Aboriginal women were not granted
the right to vote until 1960, when all Aboriginal people were finally
given the right to vote by the federal government.[1] The first women
were elected soon after women had the right to vote, but for the next

one hundred years, women would still be largely unrepresented in our parliaments.

Canada boasts the first woman elected in the British Empire, a distinction held by Louise McKinney, who won a provincial seat in Alberta in 1917, representing a left-wing socialist party called the Non Partisan League. In the same election, Roberta MacAdams, a member of the Canadian Army Medical Corps, was also elected when the overseas votes were counted. Although Alberta is not often seen as progressive in its politics relative to the rest of Canada, in 2012, Alberta elected Progressive Conservative leader Alison Redford as premier, and in 2015, Alberta had elected an NDP leader Rachel Notley, who can boast near gender parity of twenty-five female MLAs to twenty-eight male MLAS.

British Columbia was not far behind Alberta's early lead when Liberal MLA Mary Ellen Smith was elected in the January 1918 by-election, replacing her late husband. Mary Ellen Smith was a political powerhouse for years, organizing her husband's political campaigns, speaking in his place when he was not available, and fighting, successfully, for women's voting rights. She also volunteered for other causes, including the Red Cross and war veterans, and established factories to employ blind children. She founded the Liberal Laurier Club, which is an important federal Liberal association that is still active today.

Smith's slogan was "women and children first," and she stated at the time that she was proud that both women and men supported her. While part of the BC legislature, she established a minimum wage for women and children, helped establish juvenile courts, and brought in many progressive policies. She also served as a cabinet minister and Speaker of the Legislative Assembly, the first woman in either position in the British Empire.

The first female Canadian MP was Agnes Macphail, who won her seat for the United Farmers of Ontario and the CCF (the precursor to the NDP) in 1921 in the riding of York, which is Toronto. She served until 1940.

It was not until after 2010 that women were seen to be making significant political gains, for men still overwhelmingly dominated politics in terms of percentage of elected representation, cabinet positions, and leadership. In the fall of 2012, Canada's provinces finally had as many female premiers as male, and they presided over male-dominated caucuses.

When Prime Minister Justin Trudeau appointed his cabinet in the fall of 2015 and presented the same number of women and men cabinet ministers, he was repeatedly asked why he was insistent on a gender balance in his cabinet. He finally answered, "Because it's 2015."

In addition, historically, female representatives are more common in the left-wing and the moderate parties, and although there have been female leaders of right-wing parties, they have been the exception. When I was elected in 1991, I was the first pregnant woman to be elected, and one of the first to have a child while holding office. Ten years later, in 2001, Christy Clark was elected while pregnant and went straight into cabinet. And ten years later still, in 2011, Christy Clark won the by-election in Vancouver–Point Grey as premier of the province, her son Hamish at her side, breaking a thirty-year record of by-election defeats by BC's governing parties. Progress.

In the *Vancouver Sun* on May 31, 2011, Jonathan Fowlie wrote:

Christy Clark took her seat in the B.C. legislature for the first time as premier Monday morning, completing her months-long transition from the radio airwaves into the province's highest public office.

Clark returned to provincial politics with her leadership bid this past winter, and has been premier since her swearing-in ceremony on March 14.

But she couldn't sit in the legislature until after the results of her Vancouver-Point Grey by-election had become official and she was sworn in as an MLA.

That out of the way Monday morning, Clark entered the legislature to utter her first words in the house as premier.

"I am honoured to be able to introduce my son, Hamish Marissen-Clark," she said, smiling at her nine-year-old son in the public gallery before rising to excuse herself from the chamber.[2]

Christy Clark defeated three male contenders for the leadership, and she did so with almost no support from the elected caucus members. Her by-election win was narrow, and her first order of business was to expand the support base within the party and her government caucus.

Don Guy says her leadership was the opportunity the Liberal Party needed. "If Kevin Falcon had won the leadership, there's no doubt in my mind that we would have Adrian Dix as premier. I think that's one way she swayed support: all the internal polling showed she was the only one who could beat the NDP."

Campaign co-chair Sharon White says the attempts to unite everyone behind the new leader started immediately. "She had almost no caucus support, and after the leadership, she had to show that she would welcome them as part of the team. There were some significant players in caucus who stepped up to help. A few who hadn't supported her, once the numbers were announced, they were right there for us. As soon as the numbers were announced, the caucus all came up on stage to show their support."

One of the first caucus members to step up and take on a key role was Rich Coleman, MLA for Fort Langley–Aldergrove, who was considered one of the leaders of the conservative caucus members. He supported Kevin Falcon during the leadership race.

How did Coleman feel about Christy Clark's run for the leadership? "I thought it was good. I had decided that I wasn't going to run, and most of my friends who would have been working on my campaign were working on hers. It was clear it was going to be really beneficial for the party to have someone from the outside run."[3]

Coleman did not appear to have difficulty transitioning from Campbell's leadership and his position as a key supporter of Falcon's campaign. "Every leader is different," says Coleman. "The biggest change I noticed was how disciplined she was about the timing of meetings. If she had a meeting scheduled to end at 12:00, you could plan on having your next meeting start then. I noticed that she was also very disciplined about her decisions, and she wanted policy that was fiscally responsible. She was strong with her opinions; you knew where she stood on positions, and that is good."

When I embarked on this book project, I committed to be perfectly honest about my journey as a writer, and so I have to say that, of the many interviews that I did in my research, it was the interview with Rich Coleman that surprised me most. I had known of Rich Coleman since his election in 1996 and did not perceive any areas of common interest. My most difficult admission? I referred to him privately as Darth Vader. I now have to admit that I was wrong about him. After my research, hearing his passion on a number of issues and his clear indication of hard work to resolve some problems (details of which are part of a later chapter), I have completely reversed my opinion of him. In addition, his dedication to Christy Clark as leader is remarkable.

Coleman perhaps drew on some of his past training with the RCMP when Clark took over as leader. "My role was to be one of the people who had to keep it all together," says Coleman. "You just had to shelve the disloyal people to the side and tolerate them while you focused on the bigger picture. It was important to make sure there were people in the caucus who were strong and who were going to make sure no one could take down the leader or the organization, because we were not going to put up with the bad behaviour."

The tension in the caucus was high.

"The biggest adjustment with her leadership," says Coleman, "was the difficult time she had to spend with a group of people who were very disloyal. It showed me how strong a leader she was because she dealt with them with real graciousness. These were people who wanted to re-fight the leadership race. They just needed to move on so we could get on with governing. She was amazingly strong. Those of us who could see the bigger picture, stayed with her, but there were others who were really disloyal, and their behaviour reflected that."

Mike McDonald says, "Once [Christy Clark] became leader, she moved very quickly to bring the caucus together. She met with every member for at least half an hour, to speak to them directly and get to know them and what mattered to them."

According to Vaughn Palmer, the lack of caucus support "saddled Clark with some significant repair work when she did win the leadership. Plus her lone supporter, Harry Bloy, saddled Clark with a different challenge as a manager, as the record will show."

Palmer is referring to a controversy that captured Bloy about a year after Clark became premier, when Bloy was in cabinet. "Because Bloy was the only member of caucus to support Clark for the leadership, she had to put him in cabinet and did so. She ended up putting him into a ministry where he was over his head, one of the worst ministerial performances I have seen." What was worse, "he also acted as if his connection to the premier would license the exercise of bad judgment… [he] had to be demoted, then dumped from cabinet altogether."

But Bloy was the least of her problems.

"Clark, as I expect she will tell you, had a very rough ride in the caucus room," says Keith Baldrey. "It was not only the media that had doubts about her ability to win the election. Between the leadership campaign and the 2013 election there were many 'bailouts' from the caucus and announcements that MLAs would not be seeking re-election, including the two main leadership rivals, Kevin Falcon and George Abbott. Many [of the bailouts] were not surprising at all. People like Randy Hawes, John Les, Iain Black, Joan McIntyre, and Kevin Krueger were obviously no fans of Clark. And the mood within government was becoming increasingly sour and bitter as the 2013 campaign approached, so it wasn't surprising to see others bail as well."

Bruce Clark watched the stress of the tension in caucus as a political operative and as an older brother. "I have never been concerned for her," says Bruce. "There's times when people pillory her, opposing forces from every direction, and it's very hard to maintain a positive attitude, where you have to keep your spirit up and the spirit up of everyone on the team."

He saw the internal and external attacks on his sister. "People can be pretty mean … there's not a lot of accountability for the attacks," Bruce Clark says. "After things clear, they say 'that wasn't me,' but you know they were involved."

Keith Baldrey saw some of the nastiness first hand. "Things were pretty tense as I recall, and some members of that caucus wore their disdain for Clark on their sleeves. One day, I was sitting in the legislature dining room having breakfast and reading the *Province* newspaper. Several Liberal MLAs walked past and noticed the paper showed a picture of Clark praying at an Indo-Canadian temple. 'Oh look, she's praying,' one of them said, somewhat sneering. 'How appropriate.'"

The tone of the comment made an impact on Baldrey. "*Whoa!* I thought. That incident helped shape my view that things were indeed falling apart from within. I had seen that movie with the last days of the Social Credit government and the NDP government and thought we were witnessing a replay, but of course, it ended up that we weren't witnessing the same thing at all."

Politics often presents challenges to individuals that they would not experience if they remained in private life. The stakes in politics are so high: winner takes all in terms of power, the loser left to work up the ability to contest the next election, or simply fade away after the defeat. These challenges will test and sometimes build character. They also reveal character as leaders decide how to respond. Frequently, you can watch a political leader age in a short period of time as the toll of the high-stakes decisions plays out in their physical appearance.

Eighteenth-century French statesmen Guillaume-Chrétien de Malesherbes said, "We would accomplish many more things if we did not think of them as impossible." This is a good quote in terms of Clark's accomplishments, given the odds she faced as new leader of the BC Liberals, and then as premier of BC.

"People were happy that we had a new leader," says Clark, "but I knew that the caucus was going to be a really hard nut to crack. I knew exactly what that would be like. I remember taking a deep breath and thinking, *I'm happy that I won*, but I knew I would have to brace myself for some really hard work."

Luckily, McDonald says, "She built enough good will in the early days to sustain her in the tough times ahead. She was not an autocrat; she built relationships with caucus to win them over."

How tough were these times, when she had to re-enter politics, take over running the province, face off against a vocal and hostile Opposition, and deal with a caucus that didn't support her?

Says Clark, "It was the hardest thing I've ever done, going to work every day at this incredible job, where I got to make a huge difference in the world, and confronting people, about half the people at the time, who were openly hostile to me."

For most people who are dealing with stress in the workplace, there is the option of unloading some of that stress to family members or close friends. That is not an option when you are premier of the province. You have to take an oath of secrecy. You cannot even share information with your husband or wife. You certainly cannot complain about how your fellow cabinet or caucus members are behaving.

It is even worse when you are heading toward an election and any public comments about disunity or infighting are a gift for your opponents.

Clark remembers "going to them every day, especially when the house was sitting, when there is the most opportunity for mischief, and saying to them, 'Can we make it to the end of the day, or the end of the week, without an explosion?'"

It was almost dysfunctional because the negativity was not just in the caucus; it was spreading to government.

"They weren't just attacking me, they were attacking cabinet ministers and staff."

The team who helped Clark win the leadership were still in place, working to build the election machinery, and their work and commitment did not stop. This now meant expanding the team to include the MLAs who had decided to stay and run in the 2013 election. It also meant attracting new candidates in the ridings where the Liberal MLAs were not running again and finding candidates to represent the party in ridings with non-Liberal MLAs. As the Liberal Party dropped in the polls in the months leading to the 2013 election, recruitment jobs became harder.

Some of the MLAs were not disruptive; they were simply not inter-ested in working on the next election. "A lot of them," says Clark, "were just ready to leave. They were at the end of their careers. They didn't have the same investment in success that those of us who were sticking around did."

Vaughn Palmer points out that, among the MLAs, "the one guy who really went after Clark in quite a vicious way was John van Dongen. He was spreading all kinds of innuendo, but it didn't have much impact."

Palmer noted that van Dongen's biggest attack on Premier Clark was after he left her caucus in early 2012 to sit as an Independent MLA, citing her "failed leadership." He initially joined the Conservative Party of BC, but when leader John Cummins survived a leadership challenge in the fall of 2012, van Dongen left the Conservatives, too.[4]

Van Dongen wasn't finished with his attacks on Clark, even as an Independent MLA. In November 2012, he demanded an investigation into Christy Clark's potential conflict of interest in the BC Rail con-troversy. His complaint was twenty-three pages long, with forty-nine appendices, and his righteous indignation about potential wrongdo-ing was passionate.[5] The report from the conflict commissioner's office clearing Clark and her ex-husband was delivered just before the 2013 election.

In order to make sure government could operate, regardless of the political infighting, Clark hired Dan Doyle as her chief of staff. Doyle worked closely with Premier Clark and saw the impact of the caucus attacks on her.

"The one thing that describes her best in the first six months I worked for her: someone could knock her down and she would bounce back up with a smile on her face. She was getting knocked down every day, internally and externally, and she always bounced back up. That's an incredible talent."

Doyle's previous government experience was non-political, running the major projects and working on transportation infrastructure. He was plunged into the deep end of a political quagmire, and what he saw in terms of Christy Clark's response built his admiration for her.

"She's such an accomplished communicator that external blows can be countered fairly easily, but when your own people are taking shots at you, that's very hard."

Doyle agreed with Coleman's analysis of Clark's leadership style. "Our cabinet meetings start on time, everyone has a chance to say their piece, and she is a very good chairperson. She makes sure all perspectives are on the table, then she will wrap it up nicely before there is a decision. With some people, the last opinion they hear is what guides their decision. Not with her; she assimilates everything while people are talking.

"You would be surprised how bright this woman is, how visionary she is, and how tough she is. I've seen her in meetings from cab drivers to the prime minister, and she stands up for BC big time. She handles herself very well."

Doyle's respect for Clark goes beyond her work.

"I admire the way in which she manages being a mother around being a premier. I've seen her do it, and if Hamish needs anything, even if she is in a cabinet meeting, she will be there for him first. That's the right priority because, if you handle that, everything else gets done."

How did the premier maintain focus when the challenges increased? What did she say to her team?

Even when she was under attack from within her own caucus, Premier Clark said, "I never took it personally because it wasn't about me, it was about them. Which is always the way it is. I tell people who are having problems, and are under attack, whether it is bullying, or women having difficulty in the workplace, that this person is attacking you or your work because they have a problem. Don't take it personally. See it for what it is."

This perspective helped her shield herself from the worst of the personal attacks.

"Imagine you are lost in a forest," she says, "and you know that there is a town to get to and you want to get to that place, but you don't know how to get there. You have two choices. You can sit there and get nowhere at all, or you can keep moving and hope that you get somewhere.

"Every day I thought, *I'm going to keep walking. I'm going to get up every day and put one foot in front of the other. I'm not going to stop moving because I'm not going to surrender.* I knew if I surrendered, it was over."

The high stakes at play when things were bad were very clear: if she failed, the party that she had supported since she was a child, the party her father had championed when it was in the wilderness, would lose government.

"The chance for the whole election at that point was on my shoulders. If I failed, there was no time to replace me with a new leader."

Another leader might have quit, or been so overwhelmed by all the opposition and negativity that it would have been reasonable to give up and blame others. But Clark was not someone to pass the responsibility to anyone else.

"I had to survive, and our government had to survive the infighting. I would get up some days and think, *Okay, how am I going to get through the day? What's going to happen today? Okay, it's going to be a really terrible day.* And I would just say to myself, *I'm not going to keep sitting here.*"

Government cannot wait for infighting. Decisions must be made every hour, every day, just to continue the basic administration. She governed. As a new leader, she began to implement a new direction for her government. She was passionate about wanting to put a strategy in place to create jobs. That was the message she had heard over and over from people when she campaigned: *We need good jobs.* She started research for her jobs plan.

"Another thing that was hard, with respect to the division in caucus, was we were working toward our jobs plan on the policy side, and we were making real progress, but it was impossible to get half the caucus to focus on the progress we had made."

Clark was frustrated by this unexpected obstacle to governing.

The opposition to her leadership was more active than just the caucus members who were undermining her. The rumblings about the need to replace her grew as the poll numbers showed a certain defeat in the 2013 election; the mood in the BC Liberals was pretty dispirited, while

NDP supporters were excited and encouraged about their anticipated win. Certainty about the pending change in government was not simply in the elected arena. There were party members organizing for the expected loss.

"There was a big network of people who were already organized to make sure I resigned if I lost the election." Although Clark didn't take any time thinking about her options if she had lost the election, when pressed she says, "I would have done what was best for the party."

The truth was that few people, either within the party or outside of it, held out much hope of winning the 2013 election—many considered it an almost forgone conclusion that the BC Liberals' days in power were ticking down. Government staff were making plans for the change to the NDP, and these discussions happened fairly openly. My sister Josie Tyabji was one among a number of people who commented on discussions about how this was affecting meetings.

"I was chairman of the BC Wine Institute, and we had a number of policy initiatives we were working on with government staff. It was so expected that the NDP would form government after the election that they told us we had to wait on decisions because they were all focused on the transition."[6]

Christy Clark says there was a turning point in her response to the unrelenting attacks that she remembers with clarity. This was a moment that defined her and shaped her approach from that moment on, not just to the challenges she was facing, but to her choice of how to start each day. In her own words:

It was August. I was in the Gulf Islands at our family cabin, and things were looking really bad. It was 2012, and for the first time, I felt overwhelmed, thinking that I just didn't know how we were going to win the election. I was scared.

I remember swimming in the freezing cold ocean. I've swum in that ocean my entire life, and I've always been a little nervous about all the creatures in the water because, frankly, they have a tactical advantage over me because they are below me and able to see me, and I have no idea what's down there.

I had been taught to have a healthy fear of the ocean from my parents, who wanted to keep me safe. In fact, my dad and my sister almost drowned, so our family had a real emphasis on knowing that you have to have a real respect for the ocean and its dangers. So there I was, feeling scared about the election and afraid of swimming in the ocean.

I remember swimming and thinking, *You know what, I don't know if I'm going to win the election, but it all depends on me, and I'm not going to be scared about the outcome. I'm going to let go of all my fear about it. Just let it go. I'm going to take control of my feelings, and I'm going to focus.*

I am going to go after winning this election with all that I have, and I am going to jettison the fear.

And I did it. I stopped being afraid. I decided to be positive and to work hard to win. And from that moment on, I also lost my lifelong fear of swimming in the ocean.

It was a good feeling to make that decision. I told myself, *I am totally in control of my own future. I will accept the circumstances that are thrust on me, and I will make them work.*

After that decision, it was like this huge burden was lifted from my shoulders. I was able to smile and genuinely feel happy knowing that, from that point on, I would do my best, without fear.

This was the moment that I decided that *I* get to choose how I feel about every day and how I encounter it.

Ever since that day, I have gotten up and chosen to be really grateful for the day I have ahead of me.

There is a bit of irony in how some people perceive her based on her shift in perspective from this moment. She smiles when she thinks about it.

"I think that was the moment that I became the kind of optimist that people slag me about now."

Not everyone judged her negatively for her choice to be positive in the face of political challenges. She believes there were many women who watched what she endured and how she responded, and in the end they chose to support her.

"One reason that many women warmed up to me during the election is because they felt something similar to their own experience in what I was experiencing, where they were treated badly because they were ambitious, or they were underestimated because they were women."

She believes, based on her conversations with women all over the province, that there was common ground in her encounters. "What I experienced is what many women experience every day."

In the end, she takes comfort that her ability to overcome the odds was a shared victory.

"Many women said that the election proved that women could overcome huge obstacles and succeed against the odds. I especially heard from women in the South Asian community because many of them are facing double discrimination."

Her big brother understands.

"People knock her down, but she comes back up, stronger than before," says Bruce. "She's tough, tough, tough. I mean that in a good way. Behind that smile, there's no backing down."

Nothing we do, however virtuous, can be accomplished alone; therefore we are saved by love. No virtuous act is quite as virtuous from the standpoint of our friend or foe as it is from our standpoint. Therefore we must be saved by the final form of love, which is forgiveness.

DAVID BROOKS

CHAPTER 17

LEADERSHIP AND NATURAL GAS

RIGHT AWAY, Christy Clark's leadership style was noticeably different from Gordon Campbell's, and that likely expanded her base of support in caucus because she was prepared to share power.

"Night and day with Christy Clark's government," says Keith Baldrey. "She is very much a delegator. She gives her ministers enormous latitude. She doesn't pretend to try to be the forest minister, education minister, health minister, as Campbell would. She's very much a hands-off leader. Look at Steve Thomson in forestry, or Bill Bennett in energy, she lets them do their jobs."

"Ability to delegate," says Vaughn Palmer, "is one of the characteristics that most distinguishes Clark from Campbell as premier. Clark delegates very well to her senior ministers—Coleman, de Jong, et cetera, as we've seen."

Moe Sihota says her leadership style and direction is her own. "I think to a large measure, she writes her own message. And her staff may polish it. *Families first* is clearly her own message. It was rooted in her experience in politics and how she was raised as a child. This was a well-intentioned starting point, but it didn't appeal to her political base, so she drifted from it."

Post-leadership, many of the key players in the leadership race tried to bring members together under Clark's leadership, holding fundraisers and social functions, but many high-profile cabinet ministers announced that they were leaving politics. Each announcement was a blow to Clark's efforts to prop up the Liberal Party's fortunes and build for a successful election.

Says Vaughn Palmer, "Falcon organizers pushed the Tory split theme as hard as they could (during the leadership race). But he and his supporters, particularly (Ryan) Beedie and (John) Reynolds, helped the premier walk it back after the leadership. Still, Falcon decided to leave the government, as did Abbott."

She had to build on the strength of the remaining team and create a vision to attract the interest and support of the voters. Clark gave all of her ministers clear mandate letters, which set out her expectations for their ministry. Then she went to work on her key area as the new leader.

"I became premier," says Clark, "and I said we are going to have a cohesive agenda for government that applies across all ministries, and it's going to be about creating jobs." Thus began the BC Jobs Plan.

For this, she was directly engaged. "We went and dug around in government looking for ideas that would grow our economy, and we are talking about creating an environment for economic growth."

Clark recounted the steps taken, very clearly, as if she was the general entering a strategy meeting. "We went to the folks at what was then the Ministry of Energy and Mines. Someone there said, 'Well we've got this LNG thing. The previous premier wasn't interested in doing anything, but we think it's got potential.'"

And so begins the story of Liquid Natural Gas, or LNG, in BC.

"SO I'M SITTING with Pat Bell," says Clark, "who was the Minister of Jobs at the time. We're sitting there with some deputy ministers, and we are tossing around the ideas in the Jobs Plan and the deputy responsible for energy is talking about LNG."

It was an intense conversation, and Clark had to stop them.

"I said, what's LNG? And he explained it to me. He said, 'In 2005, we were talking about building an import terminal, but today, new technology has allowed us to set up the infrastructure to allow us to export it.' This idea had been kicking around in government for years, and government was just not interested in doing anything about it."

LNG allows safe and cost-effective trans-oceanic transport. Natural gas becomes a liquid when chilled to –160° Celsius (–256° Fahrenheit) and the liquid is colourless, odourless, non-corrosive, non-toxic, and it will not explode. According to the Government of Canada's Natural Resources department,

> Liquefied Natural Gas, or LNG, is simply natural gas in its liquid state... As a liquid, natural gas is reduced to one six-hundredth of its original volume, which makes it feasible to transport large volumes of gas over long distances.[1]

The new technology Clark learned about in her search for a jobs plan was the one that could transform BC's natural gas into liquefied form. Previously, BC's natural gas relied on pipelines for transport and sale, and this meant the markets had to be within Canada or the United States.

Clark was excited by what she heard: an emerging natural-gas-based energy industry would provide cleaner fuel for developing countries, create a new resource industry with good jobs in BC, and, if done responsibly, was the right way to transition away from the oil industry while creating wealth for the province and jobs for families.

"I went out and started talking to industry about what we would need to do in order to attract investment to BC for LNG." Clark leans forward and starts listing needs rapidly. "They said you need a trained work force, a competitive tax regime, good environmental legislation, you have to work with us on First Nations issues, and you have to make sure the communities are ready."

Clark leans back and smiles. "And so that is a big part of what we have been working on ever since."

Dan Doyle watched Clark take the idea forward. "She had decided she needed a jobs plan and to grow the economy, and the one resource that we have a lot of in BC, and that there was growing demand for globally, was natural gas. What she did was put the right people on the job, including Coleman, de Jong, and Bond, and they don't stop until they reach their goals."

Deputy Premier Rich Coleman has had a close view of the LNG project since its inception. "Sometimes," says Coleman, "you have people come into leadership roles at the right time and the right vision, and Christy Clark is one of them. Back in 2011, we had natural gas come forward as an idea. We were kicking the tires, and she's the one who said we are going to focus on this. We are going to pursue it. It's gutsy because LNG is not going to happen in an election cycle, so you can't get an easy political win."

Coleman understands the premier's focus and believes it's good for everyone involved.

From the government's perspective, says Coleman, "You have a resource that you can sell, a *massive* resource, and you have countries around the world that are quite literally choking on smoke from their energy sources. The right thing to do is to work for the people you represent, improve the GDP and the job situation, finance your health care and education, and supply the energy to the countries that need it. Long-term-vision-wise, it is the right thing to do. This is about the next generation of British Columbians and the generation after that. This is going to provide for my children and my children's children."[2]

Anyone who paid any attention at all to politics or government in BC since 2013 has heard about LNG, either from the advocates or the opponents. What they may not understand is how ambitious it was to try to create a new BC-based energy industry from the foundation, through to domestic procurement, meaningful jobs, and all the way to international engagement.

This goal involves almost every major issue in governance, including Indigenous rights, environmental sustainability, job creation, labour laws, tax laws, education, apprenticeship, industry partnerships, social services, procurement, trades training, global trade deals, market development, and bridging the gaps between BC's urban and rural populations so that BC can expand its ability to meet the supply challenges.

There was the long list of public policy challenges and hard work that Clark's government faced if the government took on LNG as a priority, and in addition, she had to maintain her political balance and keep the Liberals and Conservatives in her coalition happy as she communicated the development of this priority.

Clark's government's support for developing the new industry was based on the creation of high-paying, family-sustaining jobs in an energy industry that allowed a transition away from the oil sector. However, there were many critics of this approach, including wildlife consultant Guy Monty.

"I cannot see a benefit to the province from it," he said. "I can see a benefit to Liberal coffers, I can see a benefit in political currency with powerful entities around the world, but I cannot see any benefit to the people of British Columbia."

Is it all an impossible dream? Monty said the industry could be developed.

"I do believe there will be LNG plants built in BC, although I believe the number will be considerably smaller than what was discussed initially. I say there will be no benefit because of the long-term environmental costs to the people of the province and because of the ridiculous royalty system, which is a giveaway of our resources."

Jas Johal covered Christy Clark as a TV reporter for Global BC until after the 2013 election. Soon thereafter, he left his media career to work in communications for the emerging LNG industry and produced a special report on LNG in BC.

Says Johal, "All estimates tell us that energy demand in Asia will continue to grow. The key is that we have the strongest environmental standards in the world, an expensive labour force, and a lengthy consultation process, and with all of these factors, every deal is still in play. No one has left."

The scale of the projects is one reason the proposals take a long time to move forward. "One large LNG plant is $18 billion in value. It will cost more than the entire capital budget of most current infrastructure projects under way in BC."[3]

Christy Clark is not someone to shy away from hard work. Seeing LNG first and foremost as an opportunity rather than a challenge, she put LNG into her jobs plan and built it into the 2013 election campaign agenda. Communicating it would prove difficult and was complicated by the growing outcry across British Columbia against Enbridge's pipeline proposal in northern BC, the so-called Northern Gateway Pipeline.

When it came time to discuss resource issues like pipelines, social media was alive with anti-pipeline posts, and many of the participants in the discussion were dogmatically anti-pipeline, *period*. There was no room for discussion, and it didn't matter whether the proposed contents were bitumen, as in Enbridge, or natural gas, as with the proposed LNG industry.

The easiest political decision regarding Enbridge's pipeline proposal would have been to do nothing and say nothing and hope that a different political issue would take the pipeline off the front page. Environmentalists and Indigenous-rights activists were organized and growing louder, and they were strongly opposed to Clark's government. Meanwhile, Clark's business supporters and financial backers were lined up on the pro-pipeline side of Northern Gateway.

But why choose easy?

On July 24, 2012, Clark's government took a strong position on any proposed oil pipelines. Environment Minister Terry Lake and Minister of Aboriginal Relations and Reconciliation Mary Polak called a news conference on the biggest issue in BC at the time and made an announcement that was a game-changer for Northern Gateway.

BC demanded five conditions be met for all new crude-oil pipelines before they received provincial approval:

1. Completing the environmental review process, including a recommendation by the National Energy Board Joint Review Panel that the project proceed.

2. Deploying world-leading marine-response, prevention, and recovery systems for BC's coastline and ocean to manage and mitigate shipping risks and costs.
3. Using world-leading land oil-spill prevention, response, and recovery systems to manage and mitigate pipeline risks and costs.
4. Addressing Aboriginal and treaty rights and laws and ensuring First Nations had opportunities and resources to participate in and benefit from a heavy-oil project.
5. Ensuring British Columbia is paid a fair share of the economic benefits, one that reflects the risk borne by the province, the environment, and the taxpayers.[4]

Coverage of this announcement was limited, which is surprising, given that other stories on this issue were high profile and very active in terms of online comments and distribution in social media. What is remarkable is that these five conditions effectively killed the Northern Gateway proposal, an impact that was immediately recognized by Alberta Premier Alison Redford, who was critical of BC's announcement. Meanwhile, environmental groups publicly stated these conditions "did not go far enough."

Christy Clark made a number of public comments about her government's five conditions, and she became associated with the announcement. The public response was relatively accepting of the government's decision.

Not everyone believes it was a good move.

"Her position on pipelines seems to have people upset even when they are in favour of [her government]," says Guy Monty, who believes it was a political no-win decision. "She said pipelines are possible, provided you meet the five conditions. People who support the pipelines know that there is no way the conditions can be met, and people who oppose them have no trust for the government and think they are going to find a way to have them built regardless."

Since the five conditions were announced, public commentary on Northern Gateway seems to have settled down. Says Clark, "The five conditions provided an example of how, when you approach something on a matter of principle, you are far more likely to earn the acceptance of the public. People want you to do the right thing."[5]

You do not know when grace will come to you. But people who are open and sensitive to it testify that they have felt grace at the oddest and at the most needed times.

DAVID BROOKS

CHAPTER 18

THE XX FACTOR

MY RESEARCH FOR this book revealed that not only was the premier's gender a factor in her leadership, but also uncovered an "X factor," in addition to the chromosomal "xx factor," which became an "xxx factor" in some of the public commentary about her.

The *Cambridge Advanced Learner's Dictionary* defines an X factor as "a quality that you cannot describe that makes someone very special."[1] Some politicians have it, and many celebrities have it. It cannot be taught or acquired; it is an intangible *something* that catches the eye, or the attention, or the emotion of the audience.

There is an X factor to Christy Clark that is recognized, in one way or another, by almost everyone who has met or worked with her, and as a successful female player in politics, still a male-dominated arena, her chromosomes provide the XX factor, but there is a sexuality that has invaded the sexist comments about this premier on many occasions, which is an xxx factor.

As a former MLA myself, I have, of course, been long aware of the dynamics and disparities between how men and women are treated in the media. That being said, while interviewing for this book, it was actually a comment from wildlife consultant Guy Monty that convinced me to dive deeper into the particular dynamic around Premier Christy Clark.

"Working in mining and logging camps, you hear a lot of crude, sexual talk about the premier. This is bizarre to me. I'm not the kind of person who looks at every woman sexually, and with Christy Clark, I wouldn't anyway. But you are around working guys watching the news, and you hear what they are saying, and it shocks me."

Given that Monty is, by his own admission, no fan of the premier, the fact that he was offended by what he was hearing is notable. "There's something specifically about her that leads to these comments implying she would be fun at a party, but a much cruder version of that. I've been hearing this all along, during her entire public career, and it mystifies me, but it is out there."[2]

Monty's comment about the camp workers stayed with me because Monty had never met the premier—and, one assumes, neither had any of the camp workers. The truth is, however, many of the people I interviewed in the course of researching the book—including conservative, traditional, older men—described the premier as "fun," although they did so in a very respectful way, and the "fun" was always appropriate in a social context and not the key part of the interview. Another time I was shopping in a local store in Powell River and the young proprietor found out I would be talking to the premier and said, "I've always thought she would be fun to party with." This struck me as both a wonderfully human way to perceive the premier, but also very odd at the same time.

I decided I wanted an academic exploration of the topic, preferably from an outsider's point of view. I knew of a retired psychologist in Halifax who was particularly articulate in social-media discussions, so I approached Dr. Pamela Cramond-Malkin about doing a report to be included with the book. Cramond-Malkin has a doctorate in psychology from UBC and did some post-doctorate work at UCLA in neuropsychology; he was the perfect candidate to try to make sense of this XX factor without fear of any repercussions from a sitting premier. The report is presented below without commentary from me because it exceeded my expectations and deserves to stand alone.

BARBIE GOES TO VICTORIA, BY PAMELA CRAMOND-MALKIN, PHD

I am exploring the possibility of gender bias inherent in the media/ public treatment of Christy Clark as Liberal premier for the province of British Columbia. These comments are meant to explore less-often considered underlying issues, and are purposely and necessarily limited to examining power relationships generally.

Many decades past, Hedy Lamarr famously remarked: "Any girl can be glamorous. All you have to do is stand still and look stupid." Many centuries before that Petrarch wrote: "Rarely do beauty and great virtue dwell together." Both writers underscore our societal suspicion of attractiveness, especially sexiness, glamour or unusual degrees of beauty or handsomeness. We are particularly suspicious when beauty and brains coexist, and our dualistic rational brain often forces us to choose one or the other as predominant, and perhaps an oversized ladling of both beauty and intelligence seems somehow a disproportionate unfairness.

For some reason, the question often occurs, can a beautiful or sexy woman really be capable of profound thought, or is she more likely to be shallow, superficial and using her femininity to manipulate her followers?

Some feminists see all attempts to enhance or show off any beauty as pandering or acceding to a construction of male desire. Sexiness in female leaders also can be quickly identified as something deliberate, purposely provocative and needing to be separated out as a primary definer of a woman's being.

When we hear a powerful woman labeled often as a Barbie doll, a Kewpie doll, or a cheerleader, our eyes should open a little wider. When a leader is greeted in a public gathering by wolf whistles, and invitations to nude windsailing adventures, we have to know a woman is being defined as a "nice piece of ass," and not as a premier of a large Canadian province. After all, no one has historically fantasized about a roll in the hay with masculinized Margaret Thatcher, or grandmotherly, pant-suited Hilary Clinton, let alone a pinched and tired-looking Alison Redford dressed in long skirts and oxford-cloth, button-down blouses, or a serious and policy-oriented Rachel Notley. All of these women are certainly attractive, even quite pretty, but this is often superseded by an overweening dab of seriousness and competency.

If there is beauty and sexiness there, it is often consciously sublimated and any aspects of glamour are quite buried. There is often an unspoken understanding that powerful women are successful if they project matriarchal, not patriarchal forms of power.

One could contrast Clark with Angela Merkel, a quantum chemist used to cutting issues into component parts and calmly dealing with them piece by piece. Mutti is a soft-featured, maternal figure who has had success dealing with difficult male counterparts by her exercise of a certain unflappable, non-threatening, reassuring type of power and authority and is often praised for her harmonious attempts at securing solidarity slowly over time.

Comparatively, Clark might be seen as more decisive, assertive, goal directed and interpersonally transactional in her style. Is Clark really any less humble or consensus seeking than a Merkel? Clark is so often cited in commentary as "being in bed" with big industry, one would have to be blind and deaf not to recognize a uniform sexism and misogyny in such descriptors. She is almost never given the benefit of the doubt about her true motivations.

A question that often arises about Clark is her level of intelligence. Ms. Clark projects an accommodating, wide eyed graciousness, a wholesomeness, and an accessible persona by her manner and speech patterns. To those who know her behind the scenes, she is seen as whip-smart, highly informed, and no stranger to articulating complex subjects. She can think rapidly and grasp and analyze matters exceptionally well. However, what we see in an interview is a woman using monosyllabic words, short sentences, avoiding complex explanations. We see a woman raised by folks who wanted her to understand duty to her community and service. There are no pithy remarks, articulated philosophies, and a noteworthy avoidance of the complex.

This is where the dichotomous thinking of beauty or brains narrative gets legs. Such interviews, edited and parsed, are often perfect opportunities provided to her detractors to define her as shallow, or not ready to play with the big boys.

It's worth examining another political star of yesteryear who had similar characteristics to Clark, Belinda Stronach. Successful CEO of a billion-dollar business, MP Stronach was blonde, beautiful, highly intelligent, accomplished and wealthy.

In many ways, her dilemma was what Christy Clark faces. She was a little too beautiful, a little too sexy. Stronach was a strong, competent and beautiful woman seeking high office and seen as an interloper with no relevant experience, even when she had more than most male counterparts.

What followed with her emergence onto the political stage was a rapid fire and chronic, seemingly deliberate, diminishment of many impressive accomplishments in the business world. What male and female politicos talked about were her twice-divorced status and her being a single mother (but not in a good way). Her cleavage was talked about, and males often thought it was all right to say they'd like to see more of it. Her apparel choices were sometimes more important than any discussion of her policy initiatives.

What lasting impression of Stronach is left almost a decade later? Ask any male and he will still say "hot," self-seeking, opportunistic, floor-crossing "bitch." The last label thanks to the words that Peter McKay, then Minister of Defence, mouthed publicly, then denied when she dumped him.

So Stronach is the cautionary tale for Canadian women who reach too high for the brass ring of power. They risk the label of temptress, seducer, siren, and manipulator.

Research helps us here. Conspicuous good looks can be both a help and hindrance. We need only look at the experience of Prime Minister Justin Trudeau, causing hearts to flutter internationally because he is perceived by some parts of the female population as a "hottie." Physical attractiveness can elicit ogling, but it does not always translate into votes or likability. Research in recent elections in Britain and Canada suggests highly attractive candidates often get up to a 2 to 3 percent boost in electability at the ballot box. But dive a little deeper, and the analysis suggests this positive bias towards the attractive candidate is often associated with those having low political knowledge, whereas those in the know, the highly-informed and involved in politics, have the exact opposite bias.

The NDP, which seems well-blessed with thinkers, philosophers, activists, the overeducated and often underemployed, are often the pointy-headed intellectuals of the political landscape who would find Clark's lack of deep discourses in her explanations, her hotness and her (perceived) "surrender" to business interests as evidence of a lack of intelligent thinking, and she is written off as "not up to the job."

The issue of sexism and misogyny is still part and parcel of women in politics. Though much has changed in a post-modern world, and many more women seek and win high office, the issue of women in power, especially high office, causes discomfort and often engenders fear and confusion in both men and women.

Any reading of blogs or comments following news articles about Clark will often show a strong bias toward male critics who are often paternalistic and condescending in tone, often referring to her as too willing to please her corporate masters. There is often a subtext that real men of substance and self-sufficiency would know better what to do.

Any perusal of websites devoted to Christy Clark's leadership or the provincial liberal party would suggest her initiatives and accomplishments abound in a wide variety of areas. What is also evident is the assessment of any positive gains is more absent than present.

Clark is often subject to trivialization of her leadership capacities and her exercise of power is always suspected from the left, and even the centre. Commentary is often personal and vicious.

The NDP Opposition's likelihood of regaining power under Adrian Dix in the last election was seen by many NDPers as a virtual certainty. In fact, they lost, and the loss was close to shattering. One could say the aftermath was a stunned refrain of "we've been robbed" by evil corporate interests and a bimbo. Such is the extreme nature of BC politics that saying hyperbolic things about leaders and palpably despising parties with a frightening vehemence is the norm. Parliament in session is positively uncivil, and often a predictably loutish, prolonged food fight.

The narrative is that Christy Clark is so cozy with industry that she lusts at the thought of thousands of precariously-built, paper-thin oil tankers heading up and down the coast in some kind of drunken sailor stupor, and environmental disaster is a sure thing, all to make a few piddly jobs. Add to this the chronic debate about the natural gas industry as likely to pollute the aquifers, cause systemic genocide of all man or beast that might consume water, and of course cause inevitable and continuous earthquakes.

Is any of this true in the remotest of ways? The real issue, perhaps, is it doesn't matter because many people don't just believe it is true, they are instead dead certain it is the gospel truth. It's a mindset with just enough worrying facts of industry-oriented legislation and plans to make it seem partially credible to many enraged and vocal voters who often have very low political knowledge or conversely have very high levels of specific pockets of knowledge but are beset with the typical BC cognitive bias disease that seems almost genetic.

Intensely derogatory commentary, and a failure to offer credit where credit is due, is absolutely part of BC politics and makes it quite hard to sort out truth from reality.

The power Clark exercises has to be seen in light of this extreme narrative.

References to prostituting the premier's office, and even references to "skankiness," are commonplace. Opponents often resort to changing her name to "Crusty" and don't bother saying her last name.

Seldom is any mention made of Clark's competencies.

Clark is clearly gifted relationally and geared towards the empowerment of others. The groundedness and pragmatism that allow her to negotiate forcefully and well for the province's interests is almost entirely unnoticed. She is not self-aggrandizing, though speaks well of her government's accomplishments. She avoids any narcissistic spinning of her own personal accomplishments and credits others with no emotional stinginess, and she is not fixated on the perks of office.

The media, magazines, newspaper, books and the Internet at large provide constant messages about a woman's body directed at the exercised body, the stick thin body, the perky boobs body, the Kardashian-bottom body, the cosmetically tweaked face. Standards of beauty are no less important in the evaluation of a woman in high public office, with a complex set of rules in the unwritten book about beauty tips for the ascendant politician.

What would such an imaginary text say to women? Be attractive, but never sexy; wear professional, slightly masculine clothing, with well-cut lapels on toneless suits. Shoes should be comfortable and sensible; a bit of weight is all right, and a bit of underweight is all right. Bright colours and floral patterns, especially feminine ones, should be avoided.

It's all right to come to the scrum cameras looking slightly tired and a mildly disheveled, as it connotes more attention to work than to self. On the other hand, there are certain things that are positive "no nos," and they include hints of glamour, shorter, well-fitting skirts which show off well-sculpted legs, and slightly open blouses or tight upper body garments that expose cleavage even slightly. These are viewed as too sexy to allow the woman politician to be taken seriously.

Christy Clark gave a speech one day and showed a *tiny bit* of cleavage, and the whole sky nearly fell in on her. A Victorian vicar could not have become more puffed up about this outrage than the politicos across the aisle, which was characterized as if she meant to be provocative, self-promoting, and deliberately distracting, to males in particular.

The incident would be the ingredients of a skit if it were not so ridiculously puritanically hypocritical. Clark's response was to dress down even more over time. Apparently, sexiness and real beauty is not okay for women in the high offices of the land, and if you display it, you can expect gender-biased commentary, feigned outrage, and name-calling as if you were a "strumpet."

Clark handles this bit of silly theatre/morality play well. Many lesser beings might have dissolved into paroxysms of laughter. Others might have reacted with some indignation or hectored their opponents. Instead of pouting about church lady scoldings from indignant opponents, she just graciously keeps her buttons done up an inch higher.

This is the same quality of response she exhibited in the face of Richard Branson's invitation to go nude windsailing with him. It was a clinic on grace under pressure, and was pitch perfect. There was a mild rebuke, suggesting the inappropriateness of the tweeted desire, and then she moved on. Her boundaries are quietly made clear.

Years ago, Gloria Steinem, who had model good looks and could have easily inhabited the catwalks in designer shows, discomforted men and women, not just with notions of female equality, but with her beauty and her startling level of clear explanations of feminist theory. She is still despised by many to this day.

Christy Clark has class and elegance and beauty when she handles public references to her attractiveness. She takes it in her stride and her response is part of her power.

Christy Clark has had to wrestle with several high profile incidents that spring from her being an attractive woman. The Richard Branson blog post mentioned in the Cramond-Malkin report occurred in May

2012 and provoked a public response from Premier Clark in order to protect respect for her office.

"Lots of young women I hope want to run for politics," she said. "I think when you meet with the CEO of a billion dollar company who wants to do business with your province, you can get a little bit more respectful treatment than that."[3]

Branson had held a news conference the previous day about his airline's new non-stop service from Vancouver to London's Heathrow. The premier had attended, along with other VIPs. Afterward, Branson posted on his blog that Clark had accepted his invitation to go kitesurfing. A photo of a naked woman clinging to his back was included with the blog piece, and of course this post was shared all over social media. Keeping in mind that the premier is an attractive and single woman, it is a particularly inappropriate public post.

Fortunately, Premier Clark retained her sense of humour with a parting shot to Branson. "Somebody said to me, as a joke, that if that's his best pickup line then maybe there's a reason he called his company Virgin."

A few months later, another incident, this one on the radio. This time it was not a billionaire entrepreneur but a DJ for a Vancouver Island rock music station out of Courtenay. He asked the premier, on behalf of a listener, what it was like to be a MILF. Clark, caught by surprise on live radio, replied that it was better to be considered a MILF than a cougar.[4]

For those who are fortunate enough to have not heard the term MILF, it is coarse slang for a *Mother I'd Like to F*ck*. A cougar, on the other hand, is an older single woman on the prowl for younger men. The question led to weeks of controversy, and the DJ who asked the question was fired, although not at the premier's request.

The incident is just one among many examples of how this premier is often seen through a sexual lens, although the response to the Branson incident and the MILF question likely helped curtail future media commentary of her as a sexual person.

The manner of coverage of the premier has made an impact on Liberal Party president Sharon White. "To be diplomatic about the media coverage of the premier, I can say I was very disappointed. I don't believe the media has kept up with the pace of change in the province with respect to who covers politics in BC. To be candid, the reporters are all middle-aged white men, and our province is quite different from that. We have a province that is ethnically diverse with women in positions of authority, and therefore, to only have one voice means it cannot be representative of the province at large."

In one of the book interviews I had with Premier Clark, I told her that I had encountered this unusual part of the narrative, namely that she was overtly sexy and had a personal magnetism that drew people to her, and I needed to find a way to manage it in the context of her office. She looked at me in stunned disbelief, and then roared with laughter.

"If you are going to do this book, you can't be making things up!"

I told her that it was absolutely true and that, in fact, the dynamic reminded me of Bill Clinton's energy, except in a female: that everyone was drawn to her. It is much more accepted in a man, and of course is likely considered a strength in a male leader.

Vancouver Sun reporter Vaughn Palmer responded to a question about gender and Premier Clark by saying, "Now *there's* a topic for a book. Any comments on women in politics needs an asterisk: there aren't enough of them, historically, to judge in the broadest sense, the way we can with male politicians because there are so many of them."

In Palmer's career with the press gallery, he has watched women advance in leadership roles, and he says their relative rarity means "they usually get compared to male styles: sometimes to their disadvantage; sometimes in a backhanded complimentary way... See most of the comments on Margaret Thatcher. Or they get judged on the supposed advantages of women as consensus builders, more empathetic. Or we fall back on the loaded language of style, where women are judged on appearances or not allowed to get away with the aggressive style that is routine for male politicians."

Much of this, of course, mirrors the observations of Dr. Cramond-Malkin.

Says Palmer, "I think any woman who goes into a leadership position needs the confidence to establish her own brand ... so far, Clark's self-confidence on that score has been remarkable."

People who have worked closely with the premier have noticed some extreme comments she seems to attract. Says Mike McDonald, "One blogger has been so extreme that I wonder about his mental health. I decided during the leadership campaign that I wouldn't lose myself in the negativity, and I took that cue from her. She's an uplifting, positive person. It's a fact of life that many will read toxic blog posts and believe it. It happens to all leaders, but it seemed particularly rancid in Christy's case. Part sexism. Part self-interest. Part underestimating her capacities. Who knows what really drives people to hate?"

Guy Monty acknowledges that people have to pay attention to how they talk about women in leadership positions. "I have been in conversations about the difficulties that women in power face. I know that there have been a number of sexist comments made against Premier Clark. I have even made some and have been quickly reminded of this problem. I'm from a redneck background, but at the same time, I completely understand that the way we talk about women in power must be done carefully."

Mike McDonald sums it up nicely: "The narrative that she's an intellectual lightweight is a sexist narrative."

[Sorrow is] that state of mind in which our desires are fixed upon the past, without looking forward to the future, an incessant wish that something were otherwise than it has been . . . Sorrow is a kind of rust of the soul . . . and is remedied by exercise and motion.

SAMUEL JOHNSON

CHAPTER 19

DRIVING HOME

THIS IS THE part of the narrative where I, as author, enter the story of Premier Christy Clark's 2013 election campaign bid as a participant, and put the reader in the passenger seat with me as I drive home, literally and figuratively. I left the Liberal Party of BC at the convention where Gordon Campbell became leader in 1993—at that very moment, in fact—and had not followed provincial politics much for many years. Just to underscore the extent of this, I was at John Reynolds's retirement dinner reception in 2006 and saw Richard Neufeld there. I had not seen Richard for many years, but he'd been an MLA when I was an MLA. I asked him what he was doing these days.

Gordon elbowed me gently. "He's the minister of mines," he said, and gave me a pointed look.

I had to apologize profusely.

I was not really paying attention to provincial politics in 2012 when Christy Clark became leader of the BC Liberal Party. To be honest, I was angry with Christy Clark for serving as deputy premier for Gordon Campbell because I felt she had sold out. Furthermore, I still remembered how viciously she went after my husband Gordon Wilson when he was a cabinet minister in Glen Clark's government. I had appeared on her radio show once, and that reminded me how much I liked her, but her return to politics had me grumbling again. At any rate, I was busy in private business and had no time for politics, though I remember hearing a lot of negative coverage of the premier. This included news stories about the Liberal cabinet ministers who were quitting and the constant coverage of the NDP planning for government in 2013. None of this was particularly interesting to me. It was my sister Josie and my father who inadvertently planted the seeds of change in my mind.

Our family moved from Ontario to BC in 1973, when my father, Alan Tyabji, took a job with Calona Wines in Kelowna. I have three younger siblings, all sisters, and all four of us had to take turns working in the wine industry after my dad became a partner in his own winery in the mid-eighties. My sister Josie and her husband, Mike Daley, stayed in the wine industry, and at the time that Christy Clark became premier, Josie was chairman of the BC Wine Institute, which represents most of the wineries in BC, large and small. She was also a senior staff person with Constellation Brands (formerly Vincor), the largest wine company in the world. Josie was working hard on behalf of the industry, trying to bring down interprovincial trade barriers so we could ship BC wine across Canada. She also worked with the provincial government on wine taxes and licensing. Josie is not and never has been political. Like my father, she is a management accountant; they are both very business oriented.

My father, after a brief retirement, became CEO of BC Tree Fruits in Kelowna in the fall of 2012. It was in a lot of trouble when he took over, with financial issues and a need to deliver better prices for fruit to growers. As a former MLA for Kelowna, I had social-media friends who would send me information about the tree fruit industry, and I had been very worried about the orchards, the growers, and the industry since before my father took over. After my dad became CEO of BC Tree Fruits, he would share a few stories about new initiatives, and this reassured me that there was a solid change in course that would protect the growers and food production.

In order to make the necessary changes to its operations, turn the company around, and restore the commercial viability of the orchards, which would protect thousands of acres of apple and cherry trees in the

Okanagan Valley, my father and his management team needed a responsive provincial government. I didn't spend a lot of time talking to my dad about his work, but during our visits, I would hear him talk about some of the programs that were coming together. These were some of the seeds that I didn't even realize were in place until later.

Catching up with Josie over the phone one evening in late 2012, she told me that the premier had been in Kelowna for a small reception at a winery, "and she asked me if I was related to you. When I said, yes, you are my sister, she said, 'Oh, I *love* Judi Tyabji!'"

I think I growled something back over the phone like, "Well, she's a politician, what *else* is she going to say?"

"Well," said Josie, "I think she's doing a good job."

This was classic Josie, a statement of fact made in her usual accountant way. *I think she's doing a good job.* It struck me because it was not something I had heard anywhere else.

After I got off the phone with Josie, I took some time to ask myself why I was so angry with Christy Clark when she hadn't had time to define herself. I realized I was jealous. Christy Clark was someone I had known since we were teenagers. We were about the same age, and somewhere deep down, I was angry that she was premier and I wasn't.

That little bit of honesty made me laugh because I knew all too well what she had to sacrifice to be premier, and I had deliberately left politics because I was not prepared to make those sacrifices. I mentally kicked myself in the backside and decided to get over my pettiness.

Meanwhile, I was still not interested in anything political, and in our riding of Powell River–Sunshine Coast, I was planning to vote for our NDP MLA Nicholas Simons. There was a general expectation that he could be a cabinet minister after the election once the NDP won government. I still perceived Christy Clark's government to be an extension of Gordon Campbell's, and it held little interest for me.

In February of 2013, I was driving home to BC from Calgary where I had visited my daughters, Kyrie and Tanita, and my granddaughter, Maitreya. I would be stopping in the Okanagan Valley to stay with my father and stepmother in Kelowna briefly before heading west to Powell River on the northern Sunshine Coast.

It was as I drove through Kamloops to the Okanagan that the stray seeds planted by my father and Josie in the previous months began to germinate in my mind. I passed through many rural ridings, and a couple of urban ones, that had sitting Liberal MLAs. As I drove, I imposed the political noise that I had been hearing for months on top of these areas and tried to imagine scenarios where the residents would show up at the polls to vote out their Liberal MLAs in favour of NDP MLAs.

I tried hard, but I could not imagine it. Everywhere I went, I introduced casual conversations about the Clark government, waiting for the anger that I had been reading or hearing on regular or social media. I couldn't find it. Instead, I heard comments about agricultural or educational projects that were "good."

There was nothing earth-shattering or dramatic cited by people, but from gas-station attendants to grocery-store clerks to airport workers, the comments were understated and indicated an underlying calm. One worker even commented about a rise in minimum wage.

I felt like Alice in Wonderland, because even though all comments included an expectation that the NDP were going to win, and they were not unhappy about the change in government, they were also not unhappy with the current government. These comments were the opposite of everything I had been hearing and reading.

While at my dad's overnight, he told me about the major focus the government had made on turning around the tree fruit industry; a number of MLAs and cabinet ministers were working with industry leaders. It is hard to explain the relief and gratitude I felt in learning that the orchards would be saved.

Over breakfast the next morning, the kitchen television was running ads for the upcoming election, featuring NDP Leader Adrian Dix talking about BC's future. It struck me that here I was in the Okanagan, the very heart of BC, where government programs by all accounts seemed to be working and turning things around, and yet I could find almost no stories that indicated the government was doing anything productive whatsoever.

There was a big disconnect, and I was puzzled.

On the final leg of my trip home, my brain was whirring. I started to think about the various regions of the province, and imagined the next election. Images of Adrian Dix and Christy Clark played through my mind—some of them from decades past. Images of the rural parts of BC played through my mind, farms and fields, forests and mines, wineries and orchards. Eventually, I made up my mind about what I wanted to do.

I arrived home to our sheep farm in our small town in the early afternoon. I was exhausted but had to dive into farm clothes and help Gordon deliver a lamb in the field. And although this small drama drove political thoughts from my mind temporarily, as soon as we were finally inside the house and cleaned off, we sat to talk about our time apart. I didn't give him much time to contribute before I blurted out what was foremost on my mind.

"I've just come back from a long drive through the Interior, and I've decided that we need to help Christy Clark in this election. She's doing a good job as premier, even though no one is talking about it, and she deserves our support. I think she's a Liberal, and is managing to just keep things together."

To say we had a spirited discussion is a bit of an understatement. It was pretty loud at times; both Gordon and I are fairly passionate communicators when our convictions are engaged. Admittedly, Gordon had been paying more attention to politics than I had, and he was fully aware that everyone viewed Clark's government not only as a sinking ship, but one that was burning from fires set by its own crew.

He was still justifiably angry with the Liberal Party from incidents long past, and he also had absolutely no interest in re-engaging in anything political. It had been nineteen years since he lost the Liberal leadership to Gordon Campbell and almost twelve years since he had last been an MLA, and he was content to confine his public activity to his blog and some of his social-media activity. He had seen no sign that Christy Clark was trying to bring in progressive policies, and had no spare time to do anything as aggravating as involve himself in BC politics during an election.

Although I understood all of Gordon's positions, by the time I talked to him about the idea of publicly supporting Premier Clark, I was pretty firm in what I wanted to do, for my true believer had been awakened. After almost twenty years of marriage to Mr. Wilson, I knew that sometimes even a passionate "no" from him might change if I left him alone, so I stated my strong conclusion about what we needed to do and why and left him to think about it. He did what I have seen him do so many times: he went off and did some research over the next few days. And, as usual, his research was excellent.

"You're right," he said. "She has already made a number of progressive decisions, although they have received very little coverage. In addition, I've taken a look at the proposed budgets of the NDP compared to the Liberals, and there is no way we should sit idle when we might be able to help her win the election."

Gordon Wilson, as a former Minister of Finance, had been looking at the parties' financial projections. He published a blog piece that was supportive of the BC Liberal budget and critical of the NDP. "What pushed me over the edge," he says, "was the NDP's fiscal policy attacking the Liberals for tabling a 'false budget' based on unrealistic projections. Later this was proven to be unfounded. Then I heard an interview with Bruce Ralston (NDP MLA for the riding of Surrey–Whalley) on CBC where he was asked by the interviewer, 'If the NDP takes over, what do you think the debt will be?' And he said he wasn't sure how badly the Liberals had mismanaged the economy, but it would be between three and four billion dollars."

British Columbia's economic growth was faster in the years prior to 2012 than it was in the rest of Canada, private business investment was continuing to rise, and unemployment rates continuing to fall,[1] and in Wilson's view, this was a time to balance the budget. When Wilson had been finance minister with the NDP, he had overseen the creation of balanced-budget legislation; this had been repealed when Gordon Campbell became premier. When Christy Clark became Liberal Leader, she stated *repeatedly* that she was committed to balanced budgets, and she had been true to this commitment leading up to the election, tabling a balanced budget.

The NDP, in Wilson's view, were planning to be cavalier in their spending and were going to plunge BC back into a sea of red ink. Unnecessary deficit spending made him furious. "The thought that the NDP could walk in and plan to spend us into debt another four billion dollars pushed me into action."

Wilson's blog and Twitter accounts had good traction, whereas I used my Facebook page as my public forum. I told him I would write a note announcing my support for Clark over Dix, and why, and he said he wanted to write and produce a video. We did not intend to rejoin the party. In fact, Gordon and I were seen by many of the BC Liberal Party members and MLAs as the enemy because we had campaigned so openly and vocally against Gordon Campbell for so long.

Enter Mark Marissen, the compulsive power broker, intent on Christy Clark's political success. Marissen continued to support her political goals, although from the outside; he had not been welcome in the party's central campaign since Mike McDonald had told him he would threaten the viability of the coalition, which was key to any chance of electoral success.

Says Marissen, "It would have meant a big fight to insert myself into the daily workings of the central campaign because, regardless of my past political experience, I was, ironically, an outsider. Also, some people felt that I might be a polarizing influence on the dynamic of the team, and I would alienate the conservatives. There's also the awkwardness that the premier was my ex-wife. To keep the peace, I kept myself engaged in supporting the efforts via social media."

However, when the campaign approached for the election on May 14, Christy Clark insisted that Marissen be on the executive committee of the campaign, along with Stockwell Day, Sharon White, Rich Coleman, and others.

A few months before the election, Marissen created a group called BC Digital Influencers, all members of the BC Liberal Party, who were invited to join the group if they were in the top one hundred on social media with their "Klout Score." A Klout Score is created by a computer analysis of an individual's social media accounts, and the higher the score, the greater the influence the individual has through her or his posts online, whether through Facebook, Twitter, Instagram, Pinterest, or other networks.

The first group meeting was held at real estate marketer Bob Rennie's art gallery. Name tags for attendees included their Klout Score and their number of Facebook friends and Twitter followers, and it was a prestigious event, where these online commentators were able to meet the premier and meet each other in person.

Says Marissen, "[We called it] the first merit-based social media political event in world history." He laughs. "Well, it's true! We could not find any other event that was remotely similar."[2]

Paying attention to social media meant reading political Twitter and blog posts. Marissen says, "I saw Gordon Wilson defending the BC Liberal budget on Twitter. I thought *Wow, Gordon Wilson said something supportive.* I decided to try to talk to him."

In 2013, Marissen and Wilson had not met, and I had only met Marissen briefly, once, six years before. Unlike the leadership race, where Marissen was scouting Conservative credentials, now he was motivated to add some Liberal credentials. "I was getting a lot of blowback from my Liberal friends that Christy was not Liberal enough."

Marissen, still immersed in the federal Liberal Party, was surrounded by people who felt that Christy Clark was too cozy with the federal Conservatives. They wanted some indication that she was still upholding Liberal values, even as she advocated a balanced budget.

"I consider Gordon Wilson to be the founder of the modern-day BC Liberal Party," says Marissen. "Having someone like Gordon Wilson back in the fold would go a long way to showing that Christy had some real Liberal credentials that predate Gordon Campbell."

When Marissen contacted Gordon Wilson, he did it on his own initiative. He wanted to find out how Wilson felt about Clark. Marissen had no idea that Gordon and I had already decided to say something public about our support for Premier Clark.

"I thought it would be great to have some kind of public announcement that he was supporting Christy Clark," says Marissen.

But it was an uphill battle trying to convince people on the campaign team that Wilson's endorsement was a plus. If Marissen was an outsider, Wilson was seen as a wildcard and enemy. What motivation could Wilson possibly have to publicly support a losing cause?

Marissen was stubborn and determined, and he had access to the person who had the final say: the premier. "I had to fight with people about it," he says.

It was a fight he won because the premier wanted it to happen. Even with a green light from the premier, Wilson's video endorsement, its timing, release, and narrative became a very difficult project.

From where I was sitting and watching, there was a real wrestling match going on. For the first time in twenty years, Gordon and I were reconnected with people like Mike McDonald and others, and it was likely as uncomfortable for them as it was for us. Gordon had scripted a video, and each time he tried to book a time to go to the campaign office to meet with the premier or have the video filmed, the appointment was moved. We live in a small town that requires planning to travel to and from Vancouver, so as the plans became wobbly, Gordon became quite agitated.

There is no question in my mind that if Mark Marissen had not been involved in the discussions, Gordon Wilson's election campaign endorsement of Christy Clark would not have happened. Gordon would have simply walked away in frustration. It was only because Mark continued to talk with us to ensure that the endorsement moved forward that anything happened at all.

While all of this discussion and planning was going on, poll after poll showed the NDP with a large lead over the Liberals. Timing in politics is everything, and finally Gordon and I decided to organize the video ourselves and do our own distribution. Marissen negotiated the use of camera operator Marc Wang, and social-media guru Diamond Isinger was allowed to help with some informal campaign distribution of the video, but I didn't want to leave distribution and production to chance because of the tension in dealing with the central campaign.

I called Gordon's cousin Connie Wilson, who is the founder and owner of *Modern Dog* and *Modern Cat* magazines in Vancouver's Railtown, and I asked if we could use her office to shoot a video. Connie, one of the genuinely nicest people I've ever met, was happy to help. It was a short cab ride from the campaign office to *Modern Dog*, and Marissen joined us for the shoot. I staged the shot and made sure the lighting was good, then left Wilson and Marissen with the video producer, who had to time the shot so that he was not filming when any trains were running.

It was Connie's daughter Jessica's birthday, so I joined them, her sister, Jennifer, the assorted office dogs, and other staff for some cake while the video message was being filmed. We had to keep our frivolous noise to a minimum so we didn't inadvertently have "Happy Birthday" playing in the background of Gordon's political message.

Because the Liberal Party had said they did not want to be associated with the video, I set up a dedicated YouTube channel and Gmail account for it. I also contacted Ian Bailey of the *Globe and Mail* and Jas Johal of Global BC TV to give them some advance notice of the story, to see if they were interested. They were. I made sure they had advance materials so they could be ready with exclusives.

We also contacted CBC's *Early Edition*, the popular morning radio show hosted by Rick Cluff. We had an advance meeting with producer Laura Palmer so that she had an interview set for their Monday show, since the video was scheduled for release on Sunday.

Marc Wang, the camera operator and video producer, was amazing. He was able to put together both a broadcast quality and a social-media quality video in a short period of time, using considerable technical skills to stitch together separate pieces. When we returned to the campaign office to view the finished video, Mark, Gordon, and I watched it a few times, in the long and short formats, with the videographer in the large boardroom. Then it was time to wait for the premier to arrive so she could watch it and have advance warning of the message and provide private feedback.

The three of us waited in the boardroom with the laptop and a USB stick with the three video formats upon it. The boardroom was a big room with brick walls and a large table in the middle. You could probably fit twenty people around that table comfortably. The laptop was set up near the door, and Gordon sat near it as we waited. For some reason, Mark and I took seats at the far end of the table, perhaps knowing that there needed to be some space provided for Wilson and Clark to watch the video together.

We saw the premier come in and head to the main campaign war room. We waited a bit longer, and then she was with us, looking energized from the campaign, even though the headlines were still unrelentingly negative toward her. She was dressed immaculately in a fashionable pant-suit, her hair was perfect, her makeup was quietly glamorous, and she was glowing. She immediately came over and gave us a big hug, her usual way of making people feel at home. She even smelled good.

The door closed, and then Mark and I took our seats, able to hear the video but not see it. We were able to watch Premier Christy Clark as she watched the video, and I could see that Gordon was a bit wary, wondering how she would take the message. At the end, you could see she was visibly moved, and she told Gordon she was grateful for the message.

"Now my mom will vote for me," she said, and Gordon gave her a big hug. It would be two years before I would learn that in fact, her mom had passed away earlier. We had a short discussion about our plans for the video's release, including the media portion, and she thought it sounded fine.

"Next time, I'm marrying Judi," she joked. Then she was off to her next campaign stop.

When Gordon and I returned home to Powell River, I uploaded the various formats of the video to the YouTube site, keeping them unpublished until the story was out.

On May 5, the story broke with Jas Johal's piece on the Sunday evening dinner hour, the most-watched television newscast in BC.

Says Marissen, "In fact, it was a bigger impact than I ever imagined. It was like an address to the province from the premier or something. He was a major story on the 6:00 news. You can't pay for coverage like that."[3]

Wilson did what he has done before; he looked British Columbians in the eye and had a chat with them. He asked the Liberals to "come home."

"They aired the short version in its entirety," says Marissen.

Ian Bailey's *Globe and Mail* story came online the evening of May 5 with the headline "Gordon Wilson rejoins Liberal camp, backing Clark ahead of BC election" and detailed Wilson's decision. The story linked to the YouTube video. Premier Clark was quoted in the article as saying, "As a moderate voice, as someone who has been really passionately, over the last twenty years, speaking on behalf of those social issues that matter to people, I think he does have a real presence and pull."[4]

That evening, I was home alone on the farm, since Gordon had to be in Vancouver overnight to do the radio interview with Rick Cluff in studio the next morning. I had Gordon's laptop in his office upstairs, with his Twitter and Facebook accounts open, and my computer in my office downstairs with my Facebook and Twitter accounts open, and I was running between the computers while answering the phone, which was ringing off the hook. It was an extremely hot day for early May, and I remember that when I finally surfaced from the insanity, I was late to do the farm chores. When I went to the field to round up the sheep, there was a brand-new baby lamb, a beautiful black and silver female. We had never had a lamb in May before. I named her Random and had to call my neighbours to help me set up the newborn and her mum in the nursery. Social-media responses had to wait a little longer that night!

In addition to the mainstream media coverage, I made sure the You-Tube link for Gordon's video was available on all social media. It was popular, as were the news stories where they had extracted information from his address. Gordon's phone started ringing the evening the story aired, some people shocked but prepared to support him, many very angry with him for supporting Clark. Quite a bit of the media coverage was negative, including CBC's *Early Edition,* which posted, "Gordon Wilson flip flops again, this time from no political allegiances to signing on with Christy Clark."[5]

On May 8, Gordon Wilson was accorded the high profile op-ed section of the *Vancouver Sun.* He wrote it after his endorsement of Clark and was able to acknowledge a response to the negative onslaught he faced for supporting Premier Clark, plus outline the issues that brought him back into the public arena and back to the Liberal Party.[6] Key among these was Dix's announcement not to proceed with the Kinder Morgan pipeline (more on this later) and a nervousness around the NDP's fiscal plan to run high deficits and raise taxes on business.

Within days of his announcements, he was getting calls from Liberal candidates telling us that the response on the doorstep had changed and was more positive because of his announcement.

On May 8, I published a note on my Facebook page entitled "Christy Clark, Jane Sterk, or Adrian Dix? Why I choose Christy Clark," which came as a huge shock to many of my Facebook friends. My strong commitment to the environment and community-based agriculture seemed, to them, at odds with support for Christy Clark. In my note, I also outlined the reasons I could not support Adrian Dix, which went back to his time as a staff person for former premier Glen Clark. And so, Gordon and I were both on the record in 2013 as supporting Christy Clark.

Meanwhile, there were signs that voters were shifting. Campaign co-chair Sharon White says, "There were a lot of signals that showed what was going on around the province. Around the province, people were coming out to listen to [the premier], our fundraisers were sold out, and people felt motivated to show up and help out. This was quite different from what we were reading about and hearing in the media."

Mark Marissen says there was an iconic photo near the end of the campaign that he found exciting because it represented a new dynamic, one that had not been seen before. "One thing I didn't expect, which turned out to be fantastic, was the picture with Gordon Wilson and [former BC Reform leader] Jack Weisgerber, as well as Brad Bennett, which showed that Christy was actually expanding the coalition."

Three days after Gordon's video was released, a news story broke about a movement called the 801s—a group of BC Liberals sharpening their knives in anticipation of a huge loss on election night. They planned to demand Christy Clark's resignation at one minute after 8:00 on election night. Keith Baldrey and Jas Johal broke the story on Global BC. The key point was that the group contained business leaders and prominent members of the BC Liberal Party, although no one came out publicly. Buttons with the number 801 on them were created, ready to hand out on election night. News stories cited Clark's Liberal ties as part of her leadership failure.

The only person prepared to speak publicly about the need to dump Christy Clark was Martyn Brown, Gordon Campbell's former chief of staff. Global BC reported that "the movement not only wants to rid the party of Clark, but her people too, including her brother, organizers like Mark Marissen, campaign chair Mike McDonald, and the current Liberal Party president."[7]

The 801 story broke five days before the provincial election. Clark's campaign team responded with a statement that ended:

> We have growing support in every corner of the province. The party, with leaders like Stockwell Day, Jack Weisgerber, Gordon Wilson, and Brad Bennett, is working to ensure British Columbia has the strong leadership it needs to succeed.

The Liberal campaign was picking up momentum on many fronts, with campaign workers and candidates all over BC working around the clock to improve their chances. Clark's team was clearly not going to go down without a fight.

Although we returned to our small town and stayed fairly quiet after Gordon's video was released, we heard about growing optimism in the central campaign as the election came closer. The gap between the NDP and the Liberals was narrowing, although the NDP was still expected to win a majority government.

He who has a why to live for can bear almost any how.

FRIEDRICH NIETZSCHE

CHAPTER 20

SETTING
THE STAGE

C HRISTY CLARK HAS been called a polarizing leader in terms of her public image; she provokes strong responses from people, either positive or negative. What outsiders do not always see is how effectively she works behind the scenes to bring people together. She recruited political operatives and key staff people from other leadership camps, especially Kevin Falcon's, and this was a critical element in setting the stage for the pre-election campaign.

Veteran BC politician John Reynolds had supported Kevin Falcon for the leadership, and after the leadership race, there was a need to raise money. Fundraising dinners and the ability to draw in funds are absolutely essential. Reynolds said he didn't really know Christy Clark until after she was leader, and then, "I got a call asking if I would chair the dinner. They had a dinner planned, but it wasn't going very well."

Reynolds agreed and worked the phones, selling the idea of coming together under the new leader, knowing that getting people in the room was half the battle, and the other half was the leader's speech. "I co-chaired with another Conservative and the dinner sold out."

The other Conservative who helped sell out the dinner was Ryan Beedie, another high-profile former supporter of Falcon's campaign.

Reynolds said, "We did it. The party wasn't doing that well. I got on the phone and pushed a lot of people and said we cannot afford the NDP. We turned it around. Because a lot of people came out, she did a great job, going around the room. That was likely the turning point after the leadership, it made people think."

Reynolds said something that was a consistent comment about Christy Clark: "Once you get to know her, you have to like her. During and after the dinner I had a chance to get to know her."

Reynolds explains that until she became leader, "We were not in the same political party."

In Reynolds's decades of public service, he has met and worked with many leaders, from W.A.C. Bennett to Margaret Thatcher to Paul Martin to Stephen Harper. He said, "Major quality in a good leader: surround yourself with good people. She does that. Some leaders like to surround themselves with weaker people, want everyone to agree with them. A good leader makes sure she gets good people."

Dan Doyle retired from the civil service after decades of hard work. He had almost no political experience, although after leaving government he had supported Kevin Falcon's leadership bid. He was not following politics much and was surprised by a phone call at work, "I was chairman of BC Hydro, and we had a major challenge with one of the worker-safety rallies when my phone rang. I didn't recognize the number, but I answered it and heard, 'Is this Dan? This is Christy.' And it took me a minute to place the name. She asked me to come in and be the chief of staff just to get her through to the election. As soon as she asked me, I said yes."

Christy Clark had just accepted the resignation of her former chief of staff, Ken Boessenkool, who had joined her team after she won the leadership. He lasted nine months before an "incident of concern" was raised that led to his resignation. His resignation was one of a number of difficulties facing the government.

Doyle started his new position right away. "For the next week or so, every time I turned up she would say, 'Oh, good. You came back, thank God.'"

Why did he say yes in the first place? "If you have an opportunity to do a public service, you should take it. I will always say yes to an opportunity to help the government, to serve the people. At the time that I said yes, it was considered that her chances of winning the election were pretty much nil."

Dan Doyle was key to operations in the months leading to the election. "My role was to support her and her government and hold it together. You may recall we had a fractured caucus; every day, every minute, I was managing situations to try to hold the government together."

When Doyle is asked if the premier was ever vengeful or bitter after facing unrelenting, undermining attacks from her own team, he says, "The only thing she was interested in was getting to the next day, the next problem that would come along."

Says reporter Keith Baldrey, "There are many people who will view a young attractive woman as having no intellectual capacity, which is unfair, but it happens, although it can also work in a person's favour. Elections are very much about leaders. She came across as an attractive person and a positive person.

"I remember right after she won the Liberal leadership and Adrian Dix won the NDP leadership, and I was on a panel where we were discussing who had a chance to win the election. The *Province* newspaper had a big picture of Christy Clark in a hockey jersey holding a hockey stick and beaming at the camera, and they had Adrian Dix in a suit looking very uncomfortable, and I held up both pictures and said, 'Regardless of the political situation, jersey girl will beat bad-suit man every time.'"

Of course, there were many negative stories and attacks between the Liberal leadership and the election campaign. And by early 2013, Baldrey's was one of the chorus of voices predicting the near certainty of an NDP win. "I wish I had remembered that [the hockey jersey photo] when the election campaign was on, but I didn't."

Meanwhile, Christy Clark was fighting battles on all sides, and in the NDP camp, confidence was high and plans were underway for the transition to the new government. According to NDP Party President Moe

Sihota, "We were confident we would win. I thought that the caucus had built up the case for 'change' in the public's mind. I thought there was a strong appetite for change, and I also thought that the ethnic-gate scandal put the Liberals on the defensive going into the election and their brand was stained."

Sihota was referring to the ethnic outreach scandal, which generated considerable headlines in the months heading to the election. The allegation was that some government officials were spending time promoting the Liberal Party while on salary with the government and developing strategies to build support in ethnic communities in a 2012 Port Moody–Coquitlam by-election, and in the lead-up to the 2013 election.[1]

When the premier's deputy minister, John Dyble, tabled his report on the controversy, he indicated that Brian Bonney, who had been a government staff person at the time of the allegedly inappropriate behaviour, had violated some public-service standards in his actions. Bonney, who had left government for a job in the private sector, denied the allegations.

There were other allegations, later proven to be unfounded, that also generated headlines, and the NDP dubbed the story "ethnicgate." After the 2013 election, the RCMP would investigate further based on information from Adrian Dix, and there were charges laid under the Election Act of BC of inappropriate spending by the BC Liberals in the by-election.[2]

NDP Party President Moe Sihota says, "I thought [Christy Clark] had made the mistake of not calling the election early enough; as is the case when you have a new leader, there is a period of time when the Liberals were leading in the polls, and we were surprised when she didn't call the election right away."

Clark won the leadership only two years after the election, so she would have had to ignore BC's fixed election dates to call an election early. Given her long-standing advocacy of supporting fixed election dates, she would have had to violate the principle of pre-set dates and go into an election with a caucus that did not support her. She opted to govern through the remainder of the mandate.

Sihota saw a political opportunity. The NDP were watching the negative dynamic in the Liberal government and noting the frequent announcements of people quitting.

Says Sihota, "Several of her caucus members had chosen not to seek re-election. That to me gave the sense that they were about to be defeated... We thought we could prevail in that election."

Keith Baldrey points out there is one key ingredient always needed to help the NDP in BC, and that is a split in the centre-right vote. Leadership, according to Baldrey, played an important role in previous elections, including the 1991 election with new Social Credit leader Rita Johnston, BC's first female premier, and the 2001 election with new NDP leader Ujjal Dosanjh, BC's first Indo-Canadian premier.

There were, says Baldrey, "different outcomes because the new leaders, Johnston and Dosanjh, just weren't compelling figures with the public. Both the Socreds and the NDP had accumulated so much baggage, and the Socreds were impacted by the Liberals splitting the vote. Without the Liberals, it is possible that Rita Johnston and the Socreds could have won. When Dosanjh ran, they didn't have that factor, but what was unusual is that the NDP were obliterated. They were shellacked."

The BC Liberals went into the 2013 election after twelve years in government, and although there was a new leader in place, she was quite unpopular by the time the campaign began.

"No one expected the Liberals under Clark to be decimated," says Baldrey. "They thought they would be defeated. It turned out that although the Campbell Liberals had accumulated quite a bit of baggage, Clark didn't inherit that. What she did take into the campaign likely didn't matter to voters."

Jas Johal was one of the Global BC television reporters covering BC politics after Clark became leader, and he knows, from the media side, how challenging the new leader's job had to be. "She had a very hard time in the period between the leadership and the 2013 election. The Liberal caucus was leaking like a sieve."

Leaks from caucus are usually a sign that the end is near for any government, particularly for any leader, because they only happen when elected members believe they have nothing to lose. These leaks are an attempt to win favour with influential reporters in hopes that, after the government falls, the person leaking information can protect his or her own career. The most dreaded symptom is the so-called brown envelope, which usually contains confidential government documents never meant to be seen in public. They can often be quite damning.

Says Johal, "There were brown envelopes arriving to the media regularly."

A friend said it was difficult to work in an environment surrounded by negativity and constant comments about an election defeat that was just around the corner. "I never thought for one day that she was not going to win the election. It was a negative, negative, negative environment. The things people said about her were so bad, but when I met her she was so focused, so positive. I would always say to her, 'I will see you later, you are going to win.' I only ever saw her down once, when she had to fire Ken Boessenkool over sexual harassment. 'What else can go wrong?' she asked."

Meanwhile, the NDP were gearing up for the election. Moe Sihota remembers how quickly they had positive responses to their advertising campaign, which featured two people at their kitchen table trying out a new box of cereal.

"Strategically, what we did was run what turned out to be award-winning ads called 'Christy Crunch,' and they had the desired effect of saying that the Liberals may change leaders, but don't be fooled by this, it doesn't change anything. This was ahead of the curve, and the ads played very, very well."

There were many references to the BC Liberals under Christy Clark continuing the Gordon Campbell agenda, and the more the voters identified Clark with Campbell, the lower her government was in the polls.

As the government's popularity continued to erode, Keith Baldrey noted an escalation in announcements. "Prior to the election, the government was throwing Hail Mary passes all over the place because they thought they would be defeated. [Christy Clark's] opponents started to call her Premier Photo Op. She was in everything, front and centre."

There was a fairly toxic atmosphere in the social-media commentary around Premier Christy Clark, especially on Facebook and Twitter. Baldrey attributes this to the way people react to Clark. "She is more polarizing than Campbell. She seems to get under the skin of political opponents in a way that Campbell never did. Part of it is that people underestimate her. I think she likes the underdog status and she likes the comeback kid scenario. She has a youthful image, and her political opponents can't get their heads around it. Plus she's a relatively young woman, and this injects an element of sexism into the debate, even though people don't like to admit it."

A recurring iconic image of Christy Clark in the 2013 election was of the premier, in rural parts of the province, surrounded by workers, smiling, and wearing a hard hat. A number of observers commented that the story of the premier between her election as leader of the party in 2011 and the night of the 2013 election could be told by watching the evening television news with the sound switched off. She seemed to be in the heart of the resource sector or manufacturing industries, walking the plants, talking to the workers, and often dressed like them.

Jas Johal covered BC politics closely in the run-up to the election. "Christy Clark is more attuned to rural BC than many other leaders in BC. It brings to mind the years of W.A.C. Bennett and the huge infrastructure and blacktop politics of that era. She is of the same mindset, thinking big."

Although Clark is a self-described Burnaby girl who spent most of her life on the BC coast, Johal feels that the river that runs through the middle of the province, from the north and down the Interior, is the key to understanding the premier.

"The Fraser River is symbolic of who she is in BC. If you follow the river from Vancouver, and up north, where it flows, that is where her heart is."

To prepare for the election campaign, Clark needed to recruit some strong administrators to the BC Liberal Party. She also had to make sure she had people prepared to work with her on her vision for the province, or else she would be fighting another battle within her party. Clark's vision for the BC Liberal Party included ensuring that members had an opportunity to feel included in their democratic process, which meant putting people in place who opened up the party process.

Sharon White remembered the time following the leadership campaign clearly. "I was asked if I would step in to become the president of the party, which I did. This was in the spring of 2011. The party had long-term staff and an established way of doing things. I was new and had a few ideas on how to change things. I think both the premier and I have come through party politics, so we both felt at home in the party. We felt that the party can do a lot more than just show up at the elections."

It is interesting to note that when Sharon White refers to party politics, she is referring to the Liberal Party for the premier and the Conservative Party for herself. The common ground was a belief that any party functions best when the members are directly engaged in decisions.

Says White, "We wanted to make the party feel more relevant in between elections. We want it to mean something to be a member of the party and not just call on members at election time for their votes."

The timing was good for Christy Clark, for she had recruited many members during the leadership campaign. It was this dynamic that provided an opportunity to not only reshape the party but to make sure people had a chance to be involved in more than just the leadership.

"During the leadership race," says White, "there was a huge influx of members to support candidates. These were four-year memberships. The core membership is solid, and we want to make it meaningful for them to be a member."

The BC NDP elected their new leader, Adrian Dix, in April of 2011, just after Clark's win as BC Liberal leader. Dix had replaced Carole James, and was seen by most as a serious contender for premier.

I first met Adrian Dix in 1992 when I was Opposition House Leader, and Dix was working for Glen Clark, who was Government House Leader and a cabinet minister in the government of NDP premier Mike Harcourt. I also had a chance to meet Dix when my husband, Gordon Wilson, was a cabinet minister in Glen Clark's government and Dix worked as Premier Glen Clark's chief of staff. I had a close view of Dix and had actually been surprised when he won the NDP leadership in 2011 because, in my view, he seemed more an administrator than a leader.

Christy Clark had an interesting response to Dix's leadership win. "What I knew about him is that he is extremely smart; he had a reputation as a brilliant strategist."

Her knowledge of Dix's skills were based on observation when she was elected to the Official Opposition in 1996, while Glen Clark was premier. Dix earned an excellent reputation for his political skills.

Christy Clark directed her election strategy against Dix and his team on the basis of what she had seen under Glen Clark's leadership. Christy Clark, of course, had been one of Campbell's candidates in the 1996 election, when Gordon Campbell expected to win the election. Instead, Glen Clark stunned many observers by winning a second term for the NDP. Although Christy Clark was vehemently opposed to the NDP's policies when they were in government, she admired Glen Clark's skills.

"Whatever you want to say about Glen Clark, he was a masterful politician. I always assumed [Dix] was up to Glen's level. I remember thinking that they probably had chosen the best person for the job."

Given that the NDP were leading in the polls and had a new leader that Christy Clark perceived to be a brilliant strategist, she needed help. Things were not going well. The two people closest to her, Bruce Clark and Mark Marissen, were getting really worried about the team around her. They felt that she needed to bring someone from the outside who could turn the ship around.

Christy Clark was keen to expand her circle with new people and new ideas. Don Guy was one of these people. Guy, a Liberal strategist and pollster in Ontario, met her several times over the years and had a view from eastern Canada of political events in BC. He was watching Clark's career as closely as he could from that perspective.

According to Guy, "I stayed in touch with her and her team through the leadership campaign for the BC Liberal Party as that was unfolding, and after she became leader and premier, she began reaching out to me a bit more. I would offer occasional advice when asked."

What is interesting is how Guy described his entry into Clark's election machinery. "I really started to get personally motivated and engaged the dimmer it looked for her election prospects."

Based on what he knew of her leadership, he had a hard time understanding how Clark's polling numbers kept dropping. "Here's someone who is an engaging and exciting leader who represents a huge leap forward, not just for her party, but for her province and I believe just generally. And the more the media and Opposition said she couldn't be elected and were down on her, the more I paid attention to what she was saying and where she wanted to take the province. The more they got down on her, the more excited I got about her vision and her ability to win the election."

It is an irony that a veteran eastern Canadian campaigner like Don Guy could be motivated by the negativity Clark was facing on her home turf in the west. "I believed that if she could stay true to herself and communicate as effectively as only she can, then she would win the election because she would rally people to her side based on her vision, her message, and her leadership abilities."[3]

Don Guy said yes when Premier Clark asked him to come and coordinate her election campaign. "I still do occasional campaign consulting if it looks like I can make a difference and it will be fun. After a couple of long discussions, we thought I could help her succeed in the kind of campaign she would want, which would epitomize her vision, her values, and her direction."

Not many would refer to a campaign where the coverage is unrelentingly negative and the polls predict a loss in government as potentially "fun," but Guy was keen.

Don Guy became involved in the pre-election campaign and began adding to Clark's political machine. The need for good party administration extended to the staff, as well, as the party approached the 2013 election. Enter Laura Miller, a skilled Ontario Liberal organizer who was asked by Don Guy to come to BC to meet Premier Clark.

Although Miller had met Christy Clark a few times at Liberal political functions in the early 2000s, she says the first real meeting was in February of 2013, in the pre-election organizing.

"I was asked by Don Guy to come and help with the critical path of the campaign," says Miller. "It was supposed to be two days, then I met [Premier Clark], and she welcomed me to the campaign, and I was here for four months." Miller laughed at the memory of trying to negotiate with Clark. "Once she stated as a fact that I was on the team, I couldn't leave. She is very persuasive."

Laura Miller is an attractive professional who looks younger than her age. She has a wide-eyed, old-fashioned glamour to her looks, like she's walked off set from a silver-screen classic movie. She smiles easily and is a rare political operative because she does not try to dominate a conversation. I expect she wins often when playing poker; she is likely underestimated while she conceals the cards she holds. She is hard working, smart, and dedicated to causes that motivate her. Christy Clark's leadership motivated her. In 2013, she was in her mid-thirties and was best known for her excellent organizational skills and hard work in Ontario as deputy chief of staff to Liberal Premier Dalton McGuinty.

Says Miller, "My strongest memory [of the February 2013 meeting] was how welcoming she is, the familiarity, it's like she's known you and you've known her all your life. She has that ability to make you feel like you are the most important person in the world, like she cares." Miller pauses for a moment. "And she *does* care. She has such an astonishing appetite for details. She will ask about your partner, your children, even your dog; she remembers."[4]

Clark's character and her passion for BC captured Miller's imagination. "The campaign was mainly about her vision for the province and how she could win. She had such a strong vision for the province, such incredible energy and hope for its future. I felt she deserved all the support in the world. So I was happy to come and play a small role."

Miller is being modest. Her role was meant to be temporary, and instead she remained in a key staff and support position until long after the 2013 election.

The early team was being assembled before the election even as the negative commentary filled up online comments, social media, and radio talk shows. John Reynolds says that, after his decades of involvement, he believed politics in BC has always been a little like this. "People love to hate their leaders. I remember W.A.C. Bennett telling me, if there aren't people yelling and screaming outside your meetings, you're doing something wrong."

Reynolds thought Christy Clark had the necessary ingredients to do well. "You either have it or you don't have it. Christy Clark has it. No one on her team complains to me that she doesn't listen to them. She manages her team well."

IN BC, THE Indo-Canadian (or South Asian) community has played a large role in politics for decades, showing up to organize events, fundraise, and show support for leaders or political parties. There has not been strong affiliation to political parties that can be counted on from this community, which means that the same group of two hundred workers can show up for an NDP leader one month and a Liberal leader the following month, and there is often the same level of enthusiastic support. However, when the Indo-Canadian community's leadership shows up to provide high-profile exclusive support for one party or leader over another, *that* is very significant. Christy Clark benefited from such support during her leadership bid, and according to some South Asian observers, this support carried through to the election period because of her personality.

Says Jas Johal, "In the Indo-Canadian community, we tend to have pretty good radar for how comfortable politicians are with minorities. Christy Clark is very comfortable and was open with both the moderate and the orthodox Sikh members. In that way, she was very similar to [former NDP premier] Glen Clark, who also fit right in. Both good 'retail politicians.'"

The term "retail politician" is a recent phrase coined to describe a politician who enjoys meeting the voters directly, or campaigning in public areas. Prime Minister Justin Trudeau is described this way, whereas his predecessor, Prime Minister Stephen Harper, was not.

Johal is convinced that Christy Clark's popularity with and connection to the Indo-Canadian community played a strong role. "The Indo-Canadian community was key to her success. It didn't guarantee it, but it was certainly helpful. When she brought the Bollywood awards to Vancouver, that was great. She seems to understand that marketing western Canada to India is a key part of our economic diversification."

Johal is referring to Christy Clark's government's sponsorship of *The Times of India* Film Awards. It was controversial; however, BC hosted the awards ceremony in Vancouver in April of 2013, following two days of related events, including a BC–India Global Business Forum.

Jatinder Rai, a multicultural strategist, worked on the ceremony. "We had an original song written and recorded for the video to describe the immigrant experience of Indians to Canada, because music is so central to how Indo-Canadians communicate."

The song, which told the story, was called "We Left Home to Find Home." Rai says, "We went through a lot of different versions of the song to try to capture the feeling, and to include how much British Columbia has given back to immigrants."

It was a key connection for British Columbians whose origins were in India. "We made the video in three languages: Hindi, Punjabi, and English."

This was a moving tribute to the connection between these British Columbians and their ancestors' country of origin. It was a memorable audiovisual experience for attendees of the awards ceremony, and solidified the base of support that Premier Clark enjoyed, especially given her choice of honouring the community by showing up in beautiful Indian clothing when she attended the events.

As a Canadian of Indian origin, I appreciated the rich images from the whole event and thought it was an excellent way to connect to Canadians with ties to India, and to India itself.

In addition to these pockets of regional or community-based support, there were a few signs that, regardless of the polls or the negativity online, the BC Liberals still had a chance to turn their fortunes around. Mike McDonald was part of the team preparing for the election.

"We knew we were on the right track in October 2012 when we hosted our [biennial] convention in Whistler. Over a thousand BC Liberals shocked themselves by their enthusiasm and passion. It was an incredibly successful convention, as we had a lot of new blood and a genuine commitment to listening and engaging with members. They paid us back in spades by returning to their ridings fired up."

Enthusiasm in the ridings meant that volunteers would be excited about organizing and fundraising for the election and showing up when the campaign began. This energy was critical.

I firmly determined that my mannerisms and speech
in public would always reflect the cheerful certainty
of victory—that any pessimism and discouragement
I would ever feel would be reserved for my pillow.

DWIGHT EISENHOWER

CHAPTER 21

THE MACHINE
IN MOTION

PEOPLE NOT INVOLVED in politics often do not pay attention to
political leaders or parties until just before an election, or even
until just before voting day. However, for political operatives, leaders, candidates, or active members, there is considerable work to do in
the years between elections, and the pressure builds for months before
the election is called. The calling of an election is often referred to as the
"dropping of the writ," and in Canada it is triggered by the Governor General federally, or the Lieutenant-Governor provincially.

During the writ period, MLAs are released from their riding responsibilities; however, cabinet ministers and the premier still have their
responsibilities, and there is a lot of pressure on them during this
time. The hope among leaders and their ministers is that their government files are free of drama during the election so they can focus on
re-election.

The May 2013 BC election was anticipated for four years because of
BC's fixed election dates, and by February 2013, BC Liberal Leader Christy
Clark had put most of her key political operatives in place. They made up
her so-called election machine, which is critical to the opportunity for
political success. In addition to the central campaign and its staff, the

election machine for the BC Liberals included volunteers in the eighty-five provincial ridings spread throughout the province. These volunteers were responsible for local event organizing and fundraising.

Key members of Christy Clark's machine were Don Guy, Sharon White, Mike McDonald, Laura Miller, Rich Coleman, Brad Bennett (as an experienced campaigner to accompany her on the bus), and a small army of other workers at campaign headquarters. The campaign took place all over BC, with the headquarters in downtown Vancouver on the edge of Gastown.

The leader spent considerable time on a campaign bus moving around the province to greet voters and attend political events. This set-up was similar to the NDP's campaign structure. Many staff members from the political party moved into the central campaign.

President Sharon White says, "I was at campaign headquarters every day for the campaign. I was on the campaign committee, which emerged out of election readiness. As a lawyer, I also handled the legal side of the election. I stayed involved in the overall strategy for the campaign."

Mike McDonald served as campaign director in 2013. "I served as [Premier Clark's] first chief of staff following the leadership race and left the premier's office shortly after the Chilliwack by-election in May 2012. My time in government was nasty, brutish, and short. There was stress with caucus and general problems with execution; we were new. We made missteps that put us in a hole, and as time went on, expectations that we could win were very, very low."

The 2013 election was seen as an opportunity. "Having gone through the 1991 experience," says McDonald, "I was undaunted about the odds. And after I started in the party office in May 2012, Alison Redford won a shocking election victory in Alberta that confounded the pollsters and pundits. We always had a reason to hope.

"In March 2013, there were some dark moments. We had laid track for two years on jobs and the economy and on fiscal discipline. We didn't need to reinvent a platform; we carried on into the election carrying out our plan. Christy didn't micromanage; she focused on leading, doing the things only leaders can do—engaging with voters, setting a tone, and making the lonely, critical decisions that are not for the faint of heart."

The leader toured the province with a few campaign workers, a high-tech system on the bus for data exchange with headquarters and social media, and media representatives from all the major media outlets. These representatives were usually assigned to one campaign, often for most of the election, so the reporter has a chance to cover the story effectively. In BC, the election campaign is four weeks. The leader's bus tour was high profile and in the news almost every day.

Says Clark, "My first memory of the election was getting on the bus with all the media, who really appeared to be there to see the train wreck unfold."

She said the body language revealed quite a bit about what the reporters were thinking when they were assigned to spend the election campaign with her. "A lot of them would rather have been on the winner's bus. They were on the loser's bus. It was so obvious that they were kind of humouring me." She laughs at the memory. She did not let this affect her campaigning.

Up to and during the election campaign, Rich Coleman was in cabinet, which meant in addition to working on election preparation, he was working in cabinet and dealing with other members of caucus.

"I was the election readiness chair." This is the person responsible for pulling together the eighty-five candidates and their main election support. "We had some challenges in this because, obviously, we had to find new candidates, and people were all saying we were going to lose. We had to stay focused. We met regularly, and I just felt if we could get to the election day writ within fifteen to sixteen points of the NDP, we could pull it off."

Coleman never lost his faith in his leader. "I thought she could win it. There was no question in my mind that she could win the 2013 election because she is a great campaigner, and I knew we had something special in her."

The polls were dismal, and it was reasonable to expect the troops would be demoralized because everything is harder when you are entering an election where everyone believes you are a loser. It is harder to recruit candidates, bring together volunteers, and especially raise money. Elections are expensive, and fundraising is always difficult; when you are behind in the polls, it is usually almost impossible.

Surprisingly, most of Clark's key team members were confident that they could win, regardless. Party President Sharon White had to keep a positive tone as she worked with the campaign workers and membership in the spring of 2013.

"It was challenging. I am a very positive person and a true believer. I never allowed myself to think we were not going to win. It was very difficult, all the negative talk. It was hard to rally our troops. I felt that was part of my job, to keep people focused and committed. I kept our executive and others focused and motivated and continued to say, 'We are going to win.' So when you worry about money, et cetera, 'we are going to win so that will work out.'"

That's a risky strategy because, if you run a campaign anticipating a win and spend in order to win, and then you lose, it is very difficult to pay off the campaign debt. Still, White felt it was the necessary strategy for the party.

"For me, to be a true believer, I believe that what we are doing as government is the best job for British Columbia. My role as president of the party is to facilitate the environment that allows the premier to do her job. It is very important for me, as part of this community, to help make sure the place I live can be the best place possible."

Premier Clark was fortunate to have such strong support. Her supporters said she earned it. Sharon White says the premier's vision, built around a strong economy, good jobs, and balanced budgets, was part of the key to success.

"Once people heard the premier's vision for the province, we attracted some excellent candidates, people who were inspired by the premier, enough to give up successful careers in order to stand for office as candidates."

White points out that successful people do not make a decision to risk their career or public image to stand for office unless they believe that they can be elected. She said the emerging candidates were a sign that the premier's prospects were better than reported.

"If they were prepared to do this, something was happening. Around the province, people were coming out to listen to her, our fundraisers were sold out, and people felt motivated to show up and help out. This was quite different from what we were reading about and hearing in the media."

It was not just the conventional media reporting the imminent demise of Christy Clark and the impending government of Adrian Dix; social media and water-cooler talk was universally convinced that the government was about to change. For those who supported the NDP, it was a celebratory mood. There were few who would say anything positive about Clark and her government.

And yet Christy Clark stayed positive. "I knew by the time we dropped the writ," says Clark, "that I had a much better chance of winning than anyone thought. I knew that because I travelled the province and talked to people, which journalists do not do, and which pollsters do not do."

Clark enjoyed meeting people directly, and toured sawmills, wineries, and other work places around the province. She said she knew she could win "because I would go to towns across the province and people would stop me and say, 'Hey Christy how are you doing?' And they were very supportive."

This was a stark contrast to the media coverage. Says Clark, "I was a lot more hopeful about the outcome than the polls suggested I should be."

Campaign leader Don Guy had an important job, "I was to work with the premier and her team around the strategy, frame, and the message." Guy had to make sure all the core messages of the campaign were integrated across all platforms—"the advertising, leader's tour, debates, media relations, social media, etc."

A consistent message is a critical element in any campaign. It can be simple, provided it is credible and backed up by a plan. An example was

Bill Clinton's winning message, created by James Carville in 1992—"It's the economy, stupid"—designed for the campaign workers.

It is up to a good team to make sure the campaign's message is heard by the public, which happens in two key areas that the party can plan: advertising and the leaders' debates. It is up to the leader to make sure she or he carries the message through all daily campaigning because, with good co-ordination, it is repeated in the news media and social media.

Former Liberal Party president Floyd Sully was not involved in the election; however, he was a very interested observer. "What was important to me, and I think to a lot of voters, is that she stayed on message." This was repeated by many experienced political workers.

BC Liberal Party executive director Laura Miller worked on advertising and debate preparation for the 2013 election. This was a critical position because it tied in with the leader's image.

"There were usually only a few of us involved," says Miller. "I was fortunate to experience one-on-one time with [Christy Clark] as well. I got a really good sense of who she was, and what struck me is how intelligent she is—that ability to draw in all sorts of diverse information and synthesize it. There's been a lot of criticism that she's a lightweight, but in fact her capacity to read and absorb information and come up with solutions from a policy point of view is astonishing."

At that point, Miller had not worked with Clark much because her experience had been with politicians in Ontario, but certain characteristics struck her right away. "She's unbelievably direct. She doesn't sugarcoat or talk around issues. She speaks from the heart. That straight talk, that authenticity, is her biggest strength."

There are many political analysts who will say that a politician takes great risk if he or she speaks directly to an issue because there is less ability to be flexible without appearing to break a promise. Some may argue that Clark had little to lose in 2013 by taking this risk because she was so far behind in the polls. Miller said Clark's direct talk was part of why people responded.

"The reason she is able to connect with British Columbians is because she can cut through to the heart of the matter."

Christy Clark has always had solid support from her family, and perhaps no one has been more engaged in her work than her son Hamish. During the 2013 election, *Vancouver Sun* reporter Jonathan Fowlie had a chance to travel with Christy and Hamish as they began their day.

Driving across Vancouver's west side, wearing a dirt-stained Whitecaps hat, yoga pants and a black Lululemon sweater, Christy Clark is just another mother driving her son to school.

She's been on the road since 5:10 a.m., having taken eleven-year-old Hamish to an early morning goalie clinic across town. In her son's bag is the pizza and Krispy Kreme doughnut Clark packed for his lunch. Left on the dining room table at home is the raffle-ticket signup form that still needs to be completed. Just another day in the life of a working mom.

But then, from the passenger seat comes a reminder that life is different in this household, especially with a provincial election only weeks away.

"Mom, why haven't you done any political debates?" Hamish breaks the silence to ask, clearly having read the media criticism from a few days before when Clark skipped the first all-candidates debate in her Vancouver–Point Grey riding.

"Um, because there haven't been any yet," Clark responds, adjusting the knobs of her Acura SUV to clear the fog from the front windshield.

"It's coming, baby. And I'm going to debate [NDP leader] Adrian Dix as many times as I possibly—I'm going to go do as many debates as he will show up for," she says, speaking just days before she will issue a public challenge to Dix for a one-on-one debate.

"But I'm betting he's not going to show up for that many debates."

Hamish quietly hums to himself through the answer, not letting on what he thinks.

Clark says it's hard to know how much her son follows the news: "He doesn't talk about it that much."

But she knows that Hamish has an extraordinarily protective instinct.

"He knows, on every street in our neighbourhood, who had an NDP sign, who had a Green sign," she said.

She is open with him about most things, answering any questions about her job with an unflinching candour. She speaks to him several times each day, even taking his calls during meetings with senior staff.

But there are some things a mother just doesn't need to share, like the fact her son is the star of every speech she gives, in every corner of the province.

"Oh no, he doesn't know," she whispered when asked by a reporter what Hamish thinks of his regular cameos.

"I'm not worried about it," she added. "You guys never report the funny things I say anyway."[1]

Although Clark's campaign team was diverse in terms of gender and ethnicity, polling indicated that Clark's support among voters was not balanced in terms of gender.

Says Dan Doyle, "She had a hard time getting the women's vote; we did, collectively. Her most vocal and harshest critics were women, initially."

Clark did not see this reaction when she was campaigning and said that women responded to her warmly. Clark explained the response from her perspective, walking through the memory as if she was there, and explaining that the media would be at the same stop with her.

"We would make a stop, and I would walk down the street, and people would come and hug me and talk to me, and people were genuinely positive toward me, and the media completely ignored all of this." She laughed. "It was a case of cognitive dissonance. They couldn't take that information in because it didn't fit at all with their view of the world."

During the campaign, the personalities of the two leaders, Dix and Clark, started to be noticed by some of the media observers. Jas Johal of Global BC TV spent some time on the Clark campaign bus during the election and covered some of the election stories. Johal remembers thinking, *"Who would you have a drink with in the bar? Christy Clark or Adrian Dix?* The answer is clear; people enjoy spending time with Clark."

Small details sometimes had a big impact on people's responses to the leaders. "There was one story from the campaign where both leaders were asked how they eat a hamburger. Well, with Christy Clark, she just digs right in. With Adrian Dix, it was a knife and fork. All of that helped people feel that she was just like them."

The personal touch can play a big role in helping voters connect with leaders. "There was another time in the campaign," says Johal, "where we as media were filming the premier and her son Hamish called. She allowed us to keep filming while she talked to him, and she was still on camera. Dix would never allow a glimpse into his personal life; he was too uptight."

In BC politics, only one family has dominated our history for thirty of the sixty years prior to the 2013 election: the Bennett family of Kelowna. William (W.A.C.) Bennett served as BC's Social Credit premier for twenty years from 1952–72. His son, William (Bill) Bennett, served as premier for another eleven years from 1975–86. This provides a little context to the political importance of one supporter travelling on Premier Christy Clark's bus, Brad Bennett.

Brad Bennett is W.A.C. Bennett's grandson, one of Bill Bennett's four sons. He is a businessman in his early fifties who lives in Kelowna. The Okanagan Valley, in the BC Interior, was the heartland of the Social Credit dynasty. Like his father, Brad is handsome in that classic Canadian businessman way: he looks wholesome, intelligent, capable, and trustworthy. He is a family man who dresses well, works hard, and is well respected in his community. Brad Bennett also shares his family's ability to articulate an idea. He told the story of his role in working to re-elect Premier Christy Clark almost as if he was surprised by his decision to engage in politics after years of keeping a low profile.

"When I was born, my granddad was premier. W.A.C. was part of my formative years. He was premier from the time I was born until around age of fourteen. Then my dad, WR, became premier from the time I was sixteen until I was about twenty-seven. So I had no choice, I was part of a political family and politics became a part of me. You can't help but pick up things along the way and learn through osmosis."

One lesson he learned was the difference between the New Democrats and the so-called free-enterprise coalition, which was represented by the Social Credit Party. This coalition started to fall apart when Bill Bennett left politics, and after the 1991 election, the Social Credit Party never recovered its strength.

"Quite frankly," says Bennett, "I lost interest in politics when my dad retired. I took a break from it and really didn't pay attention until Gordon Campbell came on the scene and I saw it as a chance to re-engage in a new free-enterprise coalition in BC. I really became active, though, when the BC Liberals lost the 2006 election. I felt like we had missed a real opportunity to re-unite the centre-right free-enterprise coalition in BC. I took it upon myself, ever since, to remind people about the importance of staying united under a strong free-enterprise banner and never lose sight of our objective of keeping the NDP from forming government."

The work he did was behind the scenes, and Bennett resisted calls to take a prominent role. "I was helping out in local campaigns from that time on. I engaged with the premier's office from time to time, but was never heavily involved in a central campaign until the 2013 election."

Bennett resisted any involvement in the leadership race and stayed focused on the private sector in the two years between the leadership and the election campaign.

"A conversation with Mike McDonald, who was one of the key architects in putting together the election machinery, is what led to me ending up on the campaign bus. He said if the premier asked you, would you consider riding on the bus with her? I said sure, and she did. They had obviously talked it over."

Brad Bennett's commitment to help was full time, and he brought his lifetime of experience of leaders and politics, which was invaluable to the election machine. Images of him working tirelessly beside the premier provided daily credibility to her efforts. Regardless of how negative the reports were of their prospects, there was Brad Bennett working side by side with Premier Clark on all the news stories.

Brad Bennett's visible support seemed a quiet indication that there was the legacy of a provincial dynasty behind Premier Clark's efforts. "Much about campaigns today is the same as in prior elections, but much is different. Same operations, different characters, with today the added dimension of social media. Campaigns have to be run more on the clock than in my grandfather's day because of technology. He used to literally stump around the province, sometimes on the back of a train.

"I was in and out of my grandfather and father's campaigns. My memories of my grandfather's day [are] more of raucous crowds, straw hats and Dixie bands. In my dad's time it was more intense, bigger crowds. There seemed to be a competition between the parties of who could fill the bigger rooms. Today it's more about quality of information, capturing the headlines by hitting deadlines, winning online. People don't have time to sit down and watch a full news broadcast. They get their information in snippets and often on the run."

How was Premier Clark the campaigner? "There were no surprises because what she went through and how she dealt with the issues in the two years prior to the general election were not only admirable but made all of us pay more attention. I understood her strength going into the campaign, and what I saw during the campaign didn't surprise me."

This is high praise from someone as understated as Bennett is in his assessment of politics. Bennett is not a showman or someone inclined to sell his ideas. He speaks clearly about the election efforts.

"It was more of a steady push, in my opinion, through the campaign. I think most people aren't paying attention to an election or an upcoming election until the actual writ period, and then they start to pay attention to make up their minds come voting day." Bennett clearly assigns credit to Clark's leadership. "She delivered a steady, consistent message. Over time, it started to resonate. Her leadership was such a contrast to her opponent, and it became more and more obvious to voters."

It is noteworthy that, like so many others who stepped up to work with Premier Clark on the election campaign, Bennett was under no illusion that it would be easy. "When I got on the bus, at the start of the campaign, we all knew we had a tremendous fight ahead of us and a lot of work. We knew we were going to have to fight for every inch. We were polling behind fairly significantly, there's no question about that."

Bennett said, "My observation, having been around this my entire life, I don't think there's as much respect for the position of premier, the office, as there used to be. It doesn't matter whether you are a premier, MLA, MP or prime minister. I'm not sure why that degradation has taken place; it might have something to do with the intensity of media. It is too easy for us, in our offhand comments, to dismiss politicians and get away with it. It's a bit of a crowd pleaser to do that."

The premier provided a positive attitude and surrounded herself with people who could match it. "We made gains all the way through the campaign," says Bennett. "Generally we were progressing positively."

Don Guy said that it took a high level of personal commitment to be the candidate with the potential to win. "She and I share a theory on campaigning, and it is that the campaign requires almost Olympic-level style training for leaders. You have to do everything you possibly can to ensure you are in your best physical form in terms of sleeping well, exercise [for her a run every day], and keep to maximum nutrition. You can see that it worked really well."

Clark avoided alcohol during the campaign, and she watched her diet carefully to maximize her energy and her focus. Her discipline became legendary, and many of the campaign workers started talking about her high level of commitment. Frequently, this kind of leadership inspires volunteers to work even harder.

What are some of the elements of a winning campaign?

Says Jatinder Rai, "I think you do need to have a strong platform, so if she had not had the policy to back up the emotional investment, she would not have had the same success. She needed to have substance to back up the connection, and she did. Mr. Dix was different. People could not connect with him. They didn't feel a genuine life experience with him that they could connect with."

In fact, the observation that voters were having a hard time connecting to Adrian Dix was coming up more and more often as the race progressed. Increasingly Clark was seen as genuine, whereas Dix came across as aloof.

ONE OF THE key points in the campaign was the televised debate. The 2013 leaders' debate included four political leaders: John Cummins of the BC Conservatives, Jane Sterk of the Green Party, Adrian Dix of the NDP, and Christy Clark of the BC Liberals. Only the NDP and BC Liberals had elected representatives in the legislature. This showed a clear choice in the period leading up to the debate. The comparisons between Dix and Clark were made in the lead-up to the campaign, and the media were already providing analysis.

CBC ran a story saying that, strategically, the NDP's decision to release its platform in small daily doses allowed it to dominate the headlines, whereas the Liberals' decision to release the entire platform and then provide the same message every day meant their campaign was not as interesting.[2] Heading into the debate, there was a commentary in the media that Clark was having a hard time because she was left reacting to the NDP's announcements.

Predictably, the news stories following the April 30 debate reported that there had been "no knock-out punches."[3] I believe I have read that phrase in news stories after every political debate I have watched, including the 1991 televised debate in BC. Still, the Liberal campaign felt that Clark's performance had a positive impact on voters' perception of their leader.

As a political commentator, Keith Baldrey has watched decades of political debates. "Clark shone and Dix seemed to do okay, but it wasn't obvious until later that actually he bombed with many voters. It served as a reminder that the best way to watch leaders' debates is away from journalists and political aides, and sometimes with the sound down. Pictures tell the story, and which leader is strongest."

To me, Clark provided a much more credible performance, and although I was supporting her, I tried to be objective, given that I had no role in the campaign and was basically an observer.

Dix seemed uncomfortable on camera, whereas Clark's media experience came through, and she seemed to smile directly at voters through the television. Also, when Dix was asked about the memo he had created to support his former boss Glen Clark during a criminal investigation, he stated that he had been thirty-five years old. It appeared as if he was citing his age as part of his explanation, even as he took ownership of his mistake. This gave the BC Liberals an opportunity to make an online advertising campaign, where a series of their under-thirty-five-year-old candidates went on camera taking aim at Dix.

The morning after the debate, my husband and I were driving down the coast to attend a meeting in Vancouver with Mark Marissen to discuss the video project of Gordon's announcement of support for the premier. Gordon had finished scripting it, and we were still trying to get the green light from the campaign team on how to strategize the release. Gordon's cell phone rang, and I answered it, since he was driving. It was the premier, and it was the first time she and I had spoken for many years. She immediately asked how I was doing, and chatted as if she didn't have a million other pressing matters. Her warmth bounced right out of the phone and her energy level was clearly high, even the day after what is often the most stressful part of a campaign for a leader. I congratulated her on her debate performance, and she sounded genuinely surprised that I thought she had done so well. Gordon was then able to take the call; she had been calling to find out the status of the endorsement announcement.

As Christy Clark pushed hard to make headway against what seemed a guaranteed NDP win, the NDP campaign suffered another hit, this one likely more damning than Dix's debate comment. It involved the Kinder Morgan pipeline issue and efforts by the oil giant to twin an existing pipeline in order to expand their capacity to move oil through the province and to the coast, where it could expand their shipping and exports. Called the Trans Mountain Pipeline Expansion, this project proposed an expansion along the existing route, first established in 1953, from Alberta to the BC coast. Some of the expansion involved permission for new access. The issue was highly controversial in BC, and voters in the urban areas were particularly opposed to it. Burnaby, the Lower Mainland municipality that frequently voted NDP, had an active mayor who was frequently in the news assisting the anti-pipeline protest.

On Earth Day, April 22, 2013, Adrian Dix came out publicly opposing the Kinder Morgan Trans Mountain Pipeline expansion. The day after Adrian Dix opposed Kinder Morgan, NDP energy critic John Horgan called the president of Kinder Morgan. What happened during that call became a hotly debated issue.[4] The BC Liberals claimed Horgan called Kinder Morgan to indicate that, regardless of Dix's public stand, they would support the project after the election. Horgan claimed these comments by the BC Liberals were a smear campaign with no basis in reality.

The NDP, usually regarded as the political party supporting labour, workers, and well-paid jobs in the resource sector, came under fire for taking a position against Kinder Morgan's project. Union leader Tom Sigurdson cited Dix's announcement as a key turning point in the election, stating that it likely changed the vote of some of the 23,000 members of the BC and Yukon Territories Building and Construction Trades Council. Steve Hunt, Western Canada director of the United Steelworkers, believed the position was "probably a significant mistake."[5]

Says Sigurdson, "April 22, 2013, is indelibly printed in my mind. That was the day Adrian Dix announced that the party would oppose the Kinder Morgan expansion."

The frustration is still audible in his voice. "There was a process we expected the government to go through, including environmental assessment. For a political leader, whether NDP or Liberal, doesn't matter, to say they are going to do something, and decide that unilaterally without due process, that's problematic."

It was not just problematic for Sigurdson in terms of his public position, but it also caused him difficulties at work. "As soon as the announcement was made on the news, right after I watched it, my phone started ringing. People were upset. We were supporting the NDP and wanting the NDP to form government. Traditionally, that is what labour does, that's where we were aligned."[6]

Dix and Horgan's potential disagreement created an impression of confusion from the NDP. Christy Clark, like the tortoise in the old fable of the tortoise and the hare, just kept plugging away around the province with her message of jobs and the economy.

Jatinder Rai said he was watching and could feel a shift in the last two weeks. "The media polling was that we were fourteen points behind, and although I was acknowledging that this was what was reported, it didn't match what I was seeing on TV. It was all about momentum to me, and we still had twenty days left, and I could feel a steady climb."

NDP President Moe Sihota was busy in the campaign where there was every expectation of a win. "We didn't need a lot of the vote to move. With a 5 percent swing that would fracture the right-wing vote in enough ridings, this would be ample for us to win."

The NDP were hoping that the controversies surrounding Christy Clark would alienate conservative voters. "We wanted to make it unpalatable for right-wing voters to hold their nose and vote Liberal. The thought was that some would stay home, too embarrassed to vote Liberal, some would vote Conservative, and some Liberal voters would vote NDP, and this combination would be enough of a change to elect our candidates."

This was a reasonable thought, especially given that BC Conservative Leader John Cummins had been a popular and populist Member of Parliament. Former BC Liberal MLA John van Dongen had also been attacking Christy Clark's credibility for over a year with unfounded allegations connecting her to the BC Rail scandal—van Dongen crossed the floor in March of 2012 to sit as a lone BC Conservative, only to resign six months later to sit as an Independent.

Sihota talked about the strategy that the NDP adopted and how that happened. "I was part of the central campaign. Brian Topp was our campaign manager, and his gut instinct was to go for the jugular and constantly remind voters of the shortcomings of the Liberal administration and constantly make the case for change."

This approach was one discussed frequently online at the time by those who wanted the NDP to campaign against the BC Liberal government record, especially the record of the Campbell government. The phrase that Christy Clark was "Gordon Campbell in a skirt" was popular with opponents of the BC Liberal government.

The aggressive strategy was rejected by the NDP leadership. Says Sihota, "Adrian's view was that voters only needed a gentle reminder and we should stick to the positive."

"Why play nice? It is important to put this decision in context," says Sihota.

"The other variable that was very much at play is that the polling numbers from the beginning of the campaign right up until election night were telling us that we had the numbers needed to win. For those of us, me included, who wanted to make the case to go for the jugular, the numbers constrained us by showing it wasn't necessary."

NDP members around BC were excited at the prospect of the return to government, and they were counting down to election night. The campaign was tracking each day and felt that their daily announcements were succeeding at gathering a series of positive headlines and ongoing support.

For Sihota, though, the fight was not over, and he was not alone in this feeling. "There was a conversation that I recall vividly that happened on E minus 8, where Brian Topp in particular felt the need to be aggressive and go on attack ads. It corresponded with the period of time where I left the campaign office, the war room, and drove from Osoyoos to Prince George and knocked on doors, randomly, with our volunteers in the ridings. And you could discern that the Liberal vote was coming back home."

The Liberal vote "coming back home" echoes Wilson's call to "come home," and Sihota is talking about a result that was the opposite of what the NDP needed to win. They needed the vote to split.

"Brian had asked me to go do this, and we could see the gap in the numbers was closing between the parties. As a result of this exercise of talking to voters, Brian wanted to go heavily negative, but we didn't."

Sihota was a seasoned campaigner and arguably one of the most effective politicians the NDP had ever produced. His political instincts told him something was happening that was not being reported, and he agreed with Topp's push to release attack ads.

Says Sihota, "We didn't for two reasons: first, [Topp] didn't have the authority to be as aggressive as he wanted, and second because the numbers still showed us in the lead. I don't fault Brian or Adrian because the polls put us in an awkward position."

I watched the last week of the election when time permitted, following it from our Powell River home. It was impossible to tell what was going to happen, and every prediction was for an NDP win on election night. The NDP were noisy online, and the Liberal supporters were pretty quiet. I would post occasional stories of support and had made my position clear, and this led to *many* more negative comments against me than comments of support.

There was going to be a small get-together of Liberals in the area to watch the election results, and for the first time in many years, Gordon and I were invited to join a local political event. It was hosted by our mayor, Dave Formosa, who had been involved in the party since being recruited by Gordon many years before. Although we were happy to be invited, we stayed home. It was the first time in a long time that we had been so interested in the outcome of an election. Speaking for myself, I had not been so nervous and keen on a leader's victory since the 1991 election.

Gordon and I sat alone in our home and watched the election coverage, switching between the local television stations. The trend became obvious almost immediately, and the commentators all seemed to go into shock as the election outcome became clear.

In the first half hour, I started shaking Gordon's arm. (I can be very irritating when I'm excited.)

"She did it!" I kept saying.

Gordon was nervous. "Maybe, it's too soon to tell."

"She is winning!" I would say.

"We don't know yet," he said.

"She's won," I yelled.

"Okay, yes, it looks like she did." And he had a big smile on his face, and I could see he was very proud of her.

We watched every station's coverage, almost in disbelief, as the results came in. Within an hour of the polls closing, family members and friends started to call us to talk about the election.

One of the funniest parts of the evening was watching the commentators trying to find ways to interpret the results, and dealing with their reactions to a result that none of the polls had predicted.

Keith Baldrey remembers it well. "Election night itself when the voting returns shocked everyone. I was on the Global anchor desk with Chris Gailus, and we were half-wondering whether something was wrong with our computer system."

The commentators did a fair job, considering none of them was prepared for the election outcome. Also, I remember being surprised when I saw how angry Martyn Brown was. Martyn Brown, who had worked as Gordon Campbell's chief of staff, had spent a lot of time talking about the so-called 801s, the people planning to oust Christy Clark at one minute after 8:00 on election night, once she had lost the election. Brown was clearly very unhappy with Clark's victory, and that seemed odd to me, given that he was such a strong opponent of the NDP.

Even while we were feeling relieved and happy to see that Christy Clark's hard work had paid off, the NDP camp was reeling from the results.

Says Moe Sihota, "My strongest memory of the election is election night. On that night, my wife Jessie, myself, Anne McGrath, Adrian, Brian, and Stephen Howard, his aide were in a private room at the Fairmont Waterfront Hotel and we sat in stunned disappointment. Clearly we had underestimated just how resilient the free-enterprise vote is.

The NDP won three times, and each time, there was a split in the right-wing free-enterprise vote. In this instance, that split did not occur to the degree we had thought.

"It was stunning, absolutely stunning. Let's face facts: everybody's polling information was wrong, and that includes the Liberals, regardless of what they say. They did not anticipate victory that night and we did not anticipate loss."

I agree with Sihota's analysis; it was a huge surprise to everyone, even if some Liberal supporters say they were sure they could win under Clark's leadership.

Says Mike McDonald, "There were a lot of reasons why we pulled out an election victory in 2013. Boiling it down, it was similar to her leadership campaign—a lot of high-calibre people, whether they were candidates or volunteers, stepped up to commit themselves because of their support for her. They didn't do so begrudgingly; they were motivated. While everyone thought we were dead, we knew we had a very strong core to build from, and we knew that Christy had the talent, experience, and instincts to out-campaign Dix when the spotlights were the hottest."

The most quotable quote from election night? The first comment Premier Clark made when the cheers and applause finally died down while she stood smiling at the podium.

"That was easy," she said. I'm sure most of us laughed with her at the colossal understatement.

The most iconic image from the election campaign was on stage when the newly elected premier, grinning from ear to ear, shared a hug with her son, Hamish. It was a true moment.

When the stations signed off and it was all over except for the celebrations of the Liberals, the stunned aftermath for the NDP, and the questions for the media, Gordon and I turned off the TV, and it was as we were cleaning up that my cell phone rang. It was Premier Clark.

"You did it!" I yelled into her ear.

"We did it!" she yelled back.

I told her I was proud of her and hoped she would get some rest. She told me she was calling to say thank you to Gordon and to me, and that the call was the first call she had made.

"Well then go phone someone else!" I told her, laughing. "You already have our support."

Christy Clark had defied all the odds, and her machine had delivered an election victory that also elected many new members of caucus to replace the troublemakers. The 801s quietly disappeared into the shadows, and it was a night of celebration.

But Clark had a big problem. She had campaigned around the province, worked tirelessly, and delivered a majority government for her party, but she lost her seat in Vancouver–Point Grey to NDP candidate David Eby. Now she would have to fight another campaign to win a seat for herself.

We did not win it because destiny created us better than all other people. I hope that in victory that we are more grateful than proud.
ERNIE PYLE

CHAPTER 22

A NEW MANDATE

IN JANUARY OF 2013, if you told a friend, colleague, media person, commentator, or even your favourite furry friend that Christy Clark was going to win the May 2013 provincial election, the person responding would have laughed. If you were talking to your dog, it might have wagged its tail in sympathy for your delusion, and if you were talking to your cat, it would simply have stalked off in disgust. In short, you would not have had an easy time finding anyone prepared to dignify your statement with any kind of dialogue that involved Premier Clark winning the election.

And yet, on May 14, 2013, Christy Clark confounded cats, dogs, and humans when she won the fortieth general election held in British Columbia, to become the first elected female premier of the province. The Liberals won 49 of 85 seats and although winning an election is a team effort, most observers credit Christy Clark with this particular win because she didn't give up and ran a determined campaign in the face of significant internal dissent without breaking stride. Those who do not credit her for winning tend to observe that it was the NDP's election to lose, and lose they did.

So how did it happen? There are few campaigners more experienced than former MP and former MLA John Reynolds, who said he wasn't surprised that Clark won. "She stuck to the agenda, which was to talk about the economy. I remember talking to the prime minister, and he said, 'None of my BC members think she has a chance of winning,' and I said, 'That's because you are in a bubble in Ottawa.' The media coverage was all against the premier."

The voting results were the talk of the province for weeks after the election. Many political analysts tried to interpret the results, polling companies went through their methodology, and commentators were interviewed. The NDP were dealing with their unexpected loss, and so were provincial government workers.

"I think the reverberations of that loss were huge," says Moe Sihota. "I talk to bureaucrats all the time because I live in Victoria, and they were into transitional planning leading up to the election. They were getting ready for a change in government, there is no doubt about it."

What were the consequences of the election outcome? "Obviously, the election result cost Adrian his job. I think she didn't win the election; I think we lost it."

Sihota is not the only person who believes the NDP lost the election. Wildlife consultant Guy Monty says that, at first, he was surprised by the Liberal win. "Then, after thinking about it and getting over my anger, it made complete sense because the NDP chose a person to run the party who couldn't win, and they ran a really stupid campaign. They refused to engage the Liberals in what they had done wrong during the Campbell government. All they had to do was remind people what they had lost, and anyone could have won that election that way. They refused to do it, and I don't understand why."

Given that Monty is not a member of the NDP and observed the campaign from the outside, the fact that his comments match Sihota's is telling.

Says Monty, "Then, they were wishy-washy on a number of issues that matter to British Columbians. On the day of the election, I still wasn't clear on Adrian Dix's position on the Trans Mountain Pipeline, and you can't vote for someone you don't understand."

Monty spends considerable time working with people in the resource sector and socializing with people who are committed to environmental sustainability. "I don't give Ms. Clark any credit for winning that election. I'm not being small-minded or pissed off; it was the NDP screwing up. The numbers ended up being fairly close. It wasn't a landslide for Clark, and I really got the sense that it was the amalgamation of the really bad choices the NDP made at the end of the campaign so that in the polling booth, people just couldn't vote for Adrian Dix."

Sihota believed the NDP made choices in the election that left them vulnerable. "We lost it largely because we were instructed by our leader to try to avoid, as best as we can, negative attack ads on the Liberals. Adrian gave us some capacity to go after them, but he was reluctant to give us full capacity to do that. If there was a more aggressive attack on the Liberals, it would have constantly reminded the voters of the need for change. By attacking her, for those who would not vote NDP, the voters would go Conservative. By not attacking her, we allowed the Conservative voters to hold their nose and vote Liberal."

Sihota believes they have learned from 2013 and will be better positioned in the next election. "Today, the Liberals believe that they are indestructible, and with that comes some arrogance and cockiness that could be their failing in the next election."

There are, however, many who do credit Premier Clark with the win, including Brad Bennett. "She outshone her opponent, and the success of the election was all due to her."

Many leaders may not have been able to face the barrage of negativity and stay focused on the campaign work. The hard work in the campaign did yield results that the Liberal's claim they could track in the final ten days of the election.

Says Bennett, "I wasn't surprised when the final results came in because our internal pollsters were so confident, and we felt that our information was valid. I was delighted at the outcome and the fact that the premier delivered such a solid majority. It felt very good."

Don Guy claims that, on election night, "I was not surprised she won. She won because she had the most compelling vision, and as a result, she had the only message that resonated, plus she was the hardest worker in the race. She was a fabulous communicator."

Guy has strong feelings about the caucus members who bailed on Christy Clark after she became leader in 2011. "They were rats that jumped from what they thought was a sinking ship, only to watch the ship sail on to glory and leaving them swimming in the ocean. There were a lot of people who couldn't stomach a term in Opposition. It did her a favour because they took the baggage with them. It allowed her to legitimately say that it was a new party."

"I think elections are about leaders," Keith Baldrey says, "and what they project to the voters. Clark projected a positive and consistent message and she exuded confidence. By contrast, Dix was all over the map in his messaging and not very well defined either. He simply didn't look like a strong leader, and his faults were magnified when he was put next to Clark."

Baldrey also believed that the polls that were being published were either wrong, or not telling the real story about what was going on. "The polls were obviously misleading and framed a false narrative in the media. But they allowed Clark to portray herself as the comeback kid, which was an appealing image."

Vaughn Palmer put it quite simply, "She was focused, positive, and a superior campaigner, up against a guy who was weak at all those things."

Although Palmer was a seasoned veteran observer of elections, the 2013 election provided an important lesson, "My lessons learned: campaigns matter; it's not over till it's over; it's the economy, stupid; and (attention folks under the age of forty) elections are decided by the folks who actually vote."

This last comment is particularly interesting because polls frequently do not account for an age group that does not show up at the polls, so the results of the poll may be skewed. Palmer thinks one of the key factors was the "who" of the voters. "According to the data from Elections BC, turnout among registered voters skewed heavily to the older demographic in 2013. There was a 70 percent turnout in my demographic."

Palmer is a baby boomer, and clearly that is a good turnout. However, as he pointed out, older votes showed up, too. "Registered voters over the age of eighty-five voted in larger numbers than those under age forty."

How important is age in calculating voting results? Says Palmer, "In his post mortem on getting it wrong in 2013, Angus Reid said the main reason was a failure to correct his Internet panels for the prospect that younger people, who leaned NDP, don't vote. Had he made that adjustment, his election-eve poll would have pointed to a much closer result."

Direct-contact polling from a party to its supporters is usually more accurate. "[Reid's observations] dovetail with what the Liberals said about their internal surveys," says Palmer, who points out that those internal surveys "relied on telephones and doorsteps, where they got in touch with a lot of people who actually vote, and that told them they had a shot at winning if they worked harder."

His predictions for the next election? Palmer admits that, if trends hold, they favour Clark again. However, when I asked what would happen if the younger people decide to show up and vote, he was not so sure. "You raise an interesting point which will be tested in the October 19 [2015] federal election: if younger people start to engage in voting, we could see a major shift in what works on election day."

Gordon Wilson thinks there were three key factors in Clark's success. "Number one was the flip flop on Kinder Morgan by the NDP, with Horgan talking to the company and the leader saying publicly that they would not proceed. Number two was the premier's unwavering commitment to a vision of a better province. She gave people reason to hope, and that was a huge part of her success. She was tireless in her campaign to do that, and I have huge respect, having done that before. And three, I am told by people who were working in the campaign that my endorsement gave middle-of-the-road voters who were confused permission to support the party, and the support of others like Bennett and Weisgerber meant that what Christy was building was a coalition of populist ideas."

The view from the inside helped some senior campaigners to be prepared for a win. Laura Miller was not surprised that the premier won the election. "I was in the unique position, having been recruited by her and seeing her vision and how the campaign would unfold. I felt she would lead us from twenty points back to victory—and she did so marvellously."

The unique energy of the election campaign played a part, according to Miller. "In between elections, people may be dissatisfied with the government party; however, when it comes to general election campaigns, people start to pay attention to the alternatives. Her performance day in and day out, the fact that she had recruited a really strong team of candidates, a really devoted and dedicated team of political staff for the campaign, and the fact that she had such a strong vision of the future, all of this made it a winning campaign."

Premier Clark was under no illusions about the difficulty she faced in the election, and the fact that she won means she has lost some of her biggest fears. "Some reporter described my winning the last election as a 'near death experience.' There's something about facing a crisis in your life that cleanses you of fear."

A friend of Christy Clark's said the outcome of the election had a huge impact. The attacks on Clark that had been unrelenting since 2011 were finally over. "I woke up the day after the election and said, 'You have to be a grateful winner.' I had coffee with someone the next day and started to cry. I felt like I witnessed a terrible crime and now there's justice."

The consensus from most observers was that Clark outperformed Dix as leader. However, if the leader was so popular, why did she lose her riding? Mike McDonald provides some insights. "The coalition that elected us in 2013 was different from the coalition that elected us in 2009. The focus moved away from the city and university districts and toward the suburbs and the resource communities."

It is possible that this reality affected the premier's choice of ridings to represent later. "We won the same number of seats in both [elections]," says McDonald. "But the seats we lost were Point Grey, Oak Bay, and others that are similar demographics: affluent, white collar ridings with a heavy university influence. They were more federal Liberal ridings and not so [many] Conservative. The seats we gained were Delta North,

Cariboo North, Fraser–Nicola and those that benefited from the huge gains in votes from the resource sectors."

This change in support indicated a shift from the urban to the rural, and according to McDonald was significant and due to "her populist appeal and unabashed support for resource industries, especially on the Kinder Morgan issue. People with master's degrees with multimillion dollar homes in Vancouver went NDP or Green Party."

Immediately after the election, even with the government re-elected, there was still a big transition under way. Some of the cabinet ministers had not run for re-election, and the premier had to put a new cabinet together. This became a huge opportunity for Premier Clark.

Says Vaughn Palmer, "After she won, with a 50 percent turnover in the caucus room, she was able to draw on all the newcomers for genuine renewal in the cabinet room."

Still, there was a big question. Where would the premier find a riding where she could call a by-election and win a seat?

Many people do not realize that, in Canada, a leader of a political party can be premier or prime minister without being elected to the legislature. However, he or she cannot be involved in legislative debates without being elected in a riding, and it is convention for leaders to seek seats as quickly as possible.

Premier Clark had a difficult decision to make about where to run and which MLA to lose from her caucus, because someone had to resign. This is not an easy nor welcome task for any leader, and the person who resigns has to do it voluntarily.

Premier Clark was in a game of musical chairs, where one of the players had to *choose* to lose by standing up and giving her a seat, or else she could be in the impossible situation of running around with nowhere to sit. It was a very vulnerable position for her, and one that might have been much more difficult before the election.

When I asked if it was hard to find an MLA prepared to voluntarily give up his or her seat, Clark's answer surprised me. "Not at all. The strongest members of our caucus and cabinet were the first to offer. Almost every MLA came to me and offered me a chance to run in their riding."

This is an enormous sacrifice for any elected member. Even two years later, you can hear the gratitude in her voice. "It was the most amazing, affirming experience, especially after what had happened in 2011." There was no hint of the difficulties she had faced over the first two years of her leadership, and she felt she finally had a constructive group. "It was a lovely bookend to the story of how the team came together."[1]

Rumours were rampant about which riding she would choose. It seemed everyone had a theory, and since there was no news about it, people were busy making things up. My father and I were talking about it on the phone when he said that there was a credible rumour in Kelowna that the premier would choose a riding in the Okanagan Valley. *Interesting*, I thought. We talked about this for a bit. There were five ridings that our family was either living or working in, and we both agreed that it would be wonderful to have a premier from the Interior.

Gordon and I received an invitation to lunch with the premier right after the election. We were really flattered that she was taking time to meet with us, for she was likely incredibly busy with a long list of people demanding her attention. We drove down from Powell River to Vancouver to meet her on May 22. It was a hot sunny Wednesday, and the premier looked happy and relaxed as she walked out of her downtown office to greet us. We chatted about the election as we walked to the restaurant where she had booked a table. People recognized the premier and were smiling at her as we walked past. When we threaded through the tables in the restaurant, many greeted and congratulated her. She was her usual friendly self.

The premier thanked Gordon sincerely for his public support during the election, and then the food arrived. We chatted about the reaction to the election outcome and asked her where she was planning to run in the by-election. She groaned at the thought of yet another campaign, and mentioned a number of Lower Mainland seats.

"My dad says that they are expecting you to pick a seat in the Okanagan," I told her.

She looked surprised and thoughtful. "I hadn't thought about that," she said. "That's interesting."

You could see her mind whirring, weighing the options. We discussed the idea a little.

"Well, it's been a long time since we had a premier from the Interior," I said. "And the voters there are very loyal to their leaders."

The conversation moved on. Gordon and I mentioned that it was our nineteenth wedding anniversary that day.

"What are you doing to celebrate?" she asked after congratulating us.

"Having lunch with you!" Gordon said with a smile.

In truth, we had a long drive back and two ferry rides up the coast to our sheep farm after our meeting, and we were happy to make lunch the day's focal point.

In the weeks following the election, the premier was immersed in putting together her new cabinet and connecting with her team of MLAs. The cabinet swearing in was set for June 10 in Vancouver outside the Pan Pacific hotel on the waterfront.

We heard nothing more from the premier after that, other than receiving an invitation to the cabinet ceremony, along with hundreds of others. She had moved back into the high-pressure business of governing, and we were immersed in our sheep farm and business matters.

The news broke on June 5, less than a week before the swearing-in ceremony. Premier Christy Clark would be running in the central Okanagan riding of West Kelowna; MLA Ben Stewart would step aside to create the vacancy.[2] I couldn't have been more pleased. The news spread like wildfire, and apparently her choice of region caught many people completely by surprise.

Ben Stewart, owner of Quail's Gate Winery, was clearly both honoured and heartbroken to give up his seat so that the premier could contest the riding. My sister Josie, at that time Chairman of the BC Wine Institute, was surprised at the announcement and told me Stewart was well respected and many people were grateful to him for his sacrifice, saying it would be a benefit to the wine industry to have a premier with some connections in the valley.[3]

The news conference was called at a stunning location in West Kelowna, with Okanagan Lake in the background and Ben Stewart by the premier's side. Ben Stewart was a popular MLA and successful business owner. He and Brad Bennett had been friends since they were about three years old and had grown up together, so Stewart had an early exposure to politics.

Premier Christy Clark's speech recognized the Bennett legacy. "Kelowna is a natural political home for me and the values that I believe in," said Clark. "This is the cradle of free enterprise. You think of the visions that W.A.C. Bennett brought to growing this province and the vision that Bill Bennett brought to controlling government spending and keeping taxes low for the people here. I hope that with the blessing of the people from Westside–Kelowna, I can be the third premier to bring a vision to British Columbia from this community."

Clark's choice of riding took in the old Bennett homestead off Sutherland Avenue in downtown Kelowna. It was a solid choice for her, and many people showed up to help her succeed. Clark had already proved that she was a natural campaigner, and when it came to the by-election, she showed a sincere commitment to the central Okanagan, learning local issues and articulating possible solutions. Her provincial profile and the fact that she was premier meant that her attention to issues such as traffic congestion or health-care funding carried serious weight.

The campaign was in the summer, with the vote held on July 10. People from all over BC travelled to the Okanagan to volunteer, and even I made one trip to campaign for her as a former MLA from Kelowna.

Ben Stewart had won the riding in May with over 58 percent of the vote, and even though the NDP campaigned hard, Christy Clark won with 63 percent of the vote, an astonishing result given that government candidates usually do very poorly in by-elections. It was a huge vindication, and Clark's popularity in the central Okanagan was now secured. She set up a second residence there and ensured she had good staff representation.

Clark says that the view as an MLA from the Interior is quite different from that of an urban riding. "It is a very diverse province. People approach issues entirely differently depending on where they are from." Having a riding in the Interior "keeps me in touch with a key part of the province, and a premier needs to understand this region at a very personal level."

John Reynolds thinks Clark shows significant political smarts with her choices.

"It's a good area to represent... I think she could be in as long as any of the Bennetts."

TOP Premier Clark greets guests after the swearing in of her new cabinet, just after her surprise election win in 2013. Visible behind Premier Clark, from left to right, are Minister of Education Peter Fassbender; Minister of Health Terry Lake; Minister of Forests Steve Thomson; (in front) Minister of Jobs, Tourism, and Skills Training Shirley Bond; and Minister of Environment Mary Polak.

BOTTOM Premier Clark and Japanese Princess Takamodo in June 2013 continue discussions about economic development that began during the BC Trade Mission to Japan in May 2012.

TOP Premier Clark joins members of the Truth and Reconciliation Commission, church leaders, and Aboriginal organizations as they open the sixth national TRC event in Vancouver. COURTESY OF THE GOVERNMENT OF BRITISH COLUMBIA

BOTTOM Premier Clark signs a new three-year action plan with Governor Kim Moon-Soo, and celebrates BC's sister province relationship with Korea's Gyeonggi Province. COURTESY OF THE GOVERNMENT OF BRITISH COLUMBIA

TOP Premier Clark, local community leaders, Kelowna mayor Colin Basran (to the right of Premier Clark), and Kelowna MLAs Norm Letnick (second from left) and Steve Thomson (second from right) join in the groundbreaking of Pleasantvale, a housing complex for seniors and families with low to moderate incomes. COURTESY OF THE GOVERNMENT OF BRITISH COLUMBIA

BOTTOM Premier Clark attends the keynote luncheon at the annual LNG in BC Conference in 2014 and is joined by Marvin E. Odum, president of Shell Oil, in a discussion moderated by Jas Johal. COURTESY OF THE GOVERNMENT OF BRITISH COLUMBIA

ABOVE Canucks president Trevor Linden presents Premier Clark with a custom jersey in Prince George at the start of the Canucks training camp in 2015.
COURTESY OF THE GOVERNMENT OF BRITISH COLUMBIA

OPPOSITE, TOP Premier Clark helps cut the ribbon at the Conifex Biomas Opening in Mackenzie, BC, expanding the jobs at a site that has introduced innovative energy production. COURTESY OF THE GOVERNMENT OF BRITISH COLUMBIA

OPPOSITE, BOTTOM Premier Clark provides a keynote speech at the annual LNG in BC conference in Vancouver in 2015, focusing on industry best practices and global leadership in responsible natural gas production. COURTESY OF THE GOVERNMENT OF BRITISH COLUMBIA

TOP Premier Clark tours Diacarbon Biofuels in Merritt, BC, where technology is being applied to create carbon-neutral and carbon-negative biofuels. COURTESY OF THE GOVERNMENT OF BRITISH COLUMBIA

BOTTOM Premier Clark, on her third trade mission to China in November 2015, meets with China National Petroleum Corporation to focus on clean energy to supply China's growing energy demands. COURTESY OF THE GOVERNMENT OF BRITISH COLUMBIA

TOP Premier Clark provides a keynote speech about BC's best practices in responsible production of natural gas instead of coal-generated power to meet China's growing need for energy. COURTESY OF THE GOVERNMENT OF BRITISH COLUMBIA

BOTTOM Premier Clark tours Fortis BC's Tilbury LNG expansion in 2015, which created expanded employment, apprenticeship, and First Nations training programs. COURTESY OF THE GOVERNMENT OF BRITISH COLUMBIA

TOP Premier Clark visits with students, teachers, and administrators at Haida Gwaii to discuss the need for a school renovation project. COURTESY OF THE GOVERNMENT OF BRITISH COLUMBIA

BOTTOM In late November 2015, Premier Clark meets with Prime Minister Trudeau to discuss BC's program of carbon credits in advance of COP21, the global climate change conference in Paris. COURTESY OF THE GOVERNMENT OF BRITISH COLUMBIA

TOP CBC's Rosemary Barton is shown on the big screen asking questions of a panel of Canadian first ministers at COP21 in Paris. From left to right: Alberta premier Rachel Notley, BC premier Christy Clark, Ontario premier Kathleen Wynne, Prime Minister Justin Trudeau, Quebec premier Philippe Couillard, and Saskatchewan premier Brad Wall. COURTESY OF THE GOVERNMENT OF BRITISH COLUMBIA

BOTTOM Premier Clark visits the Village at Smith Creek, a seniors' home in her riding of West Kelowna, and speaks with resident Robert Scowcroft. COURTESY OF THE GOVERNMENT OF BRITISH COLUMBIA

TOP, LEFT Premier Clark meets an enthusiastic puppy as as she announces changes to the Guide and Service Dog Act: higher training standards, improved accessibility, and strengthened public safety. COURTESY OF THE GOVERNMENT OF BRITISH COLUMBIA

TOP, RIGHT The Great Bear Rainforest is announced on February 1, 2016: an ecosystem-based management plan for protected coastal areas. COURTESY OF THE GOVERNMENT OF BRITISH COLUMBIA

BOTTOM, LEFT Premier Clark greets a First Nations child at the Great Bear Rainforest announcement. COURTESY OF THE GOVERNMENT OF BRITISH COLUMBIA

BOTTOM, RIGHT Premier Clark and international music star Michael Bublé announce a $15 million grant to create the BC Music Fund to support music tourism and talent. COURTESY OF THE GOVERNMENT OF BRITISH COLUMBIA

TOP Premier Clark; BC International Trade Minister Teresa Wat; BC Minister of Justice and Attorney General Susan Anton; federal Minister of Justice Jody Wilson-Raybould; Mayor Gregor Robertson; Parliamentary Secretary John Yap; Minister of Technology, Innovation and Citizens' Services Amrik Virk; and others celebrate Vancouver's 43rd annual Chinatown Spring Festival Parade on February 14, 2016. COURTESY OF THE GOVERNMENT OF BRITISH COLUMBIA

BOTTOM Kelowna's Save-On-Foods wine store is the first of its kind in BC, and offers only BC-made beverages. To the left of the premier are Minister of Agriculture Norm Letnick and Minister of Forests Steve Thomson. (The author's sister Josie Tyabji is just visible in the crowd.) COURTESY OF THE GOVERNMENT OF BRITISH COLUMBIA

TOP In February 2016 Premier Clark addresses the Vancouver Board of Trade to discuss the balanced budget and some of its key spending. COURTESY OF THE GOVERNMENT OF BRITISH COLUMBIA

BOTTOM Premier Clark addresses the Kelowna Chamber of Commerce in 2016 to discuss her fourth balanced budget and BC's strong economic performance. COURTESY OF THE GOVERNMENT OF BRITISH COLUMBIA

TOP, LEFT On March 2, 2016, Premier Clark addresses the Globe Summit, a biennial conference in Vancouver that brings together international leaders in sustainable business. COURTESY OF THE GOVERNMENT OF BRITISH COLUMBIA

TOP, RIGHT Premier Clark receives a gift paddle from Chief Don Roberts of the Kitsumkalum of the Tsimshian First Nation, when the province delivers $9 million for skills training and socio-economic development for the Tsimshian. COURTESY OF THE GOVERNMENT OF BRITISH COLUMBIA

BOTTOM, LEFT Premier Clark embraces a Syrian refugee boy who has arrived safely in Prince George in March 2016. COURTESY OF THE GOVERNMENT OF BRITISH COLUMBIA

BOTTOM, RIGHT Premier Clark receives expert advice on premium wood crafting from famous *Timber Kings* TV show stars and expert pioneer log home builders Joel Roorda (left) and Bryan Reid, Jr. (right), in Williams Lake. MLA Donna Barnett can be seen to the left in the back. COURTESY OF THE GOVERNMENT OF BRITISH COLUMBIA

TOP, LEFT Premier Clark poses in one of several photos at the BC Muslim Association Women's Gala in April 2016. COURTESY OF THE GOVERNMENT OF BRITISH COLUMBIA

TOP, RIGHT Premier Clark talks with police officers after announcing an expansion of funding for BC's Guns and Gangs Strategy by $23 million, which included new initiatives for safer communities. COURTESY OF THE GOVERNMENT OF BRITISH COLUMBIA

BOTTOM Premier Clark has a private moment with a young woman during a day of Vaisakhi festivities on the Lower Mainland in 2016. COURTESY OF THE GOVERNMENT OF BRITISH COLUMBIA

Usage of words like 'character,' 'conscience,' and 'virtue' all declined over the course of the twentieth century. Usage of the word 'bravery' has declined by 66 percent over the course of the twentieth century. 'Gratitude' is down 49 percent. 'Humbleness' is down 52 percent and 'kindness' is down 56 percent.

DAVID BROOKS

CHAPTER 23

FIRST NATIONS VOICES

IT CAN BE reasonably argued that Christy Clark's government's greatest conflicts, and accomplishments, are both with BC's First Nations. News stories report successful economic deals made with First Nations almost as often as they show us road blockades and protests. We read about treaty talks, resource plans, and serious issues with children in care, but there is often little context. The evolving partnership with BC First Nations lies at the heart of Christy Clark's agenda and mandate. I interviewed some First Nations people, and without exception they provided intelligent, insightful, wise, and deep input into complicated issues with a focused passion that was impressive.

There are three primary organizations active for First Nations in BC: the Union of BC Indian Chiefs (UBCIC), the First Nations Summit, and the BC Assembly of First Nations (BCAFN). Each has a different mandate.

The UBCIC says the rights of First Nations have not been extinguished; therefore, Aboriginal laws have precedence over any other laws.[1] The First Nations Summit negotiates with the provincial and federal government on land title, treaty issues, and areas of mutual concern.[2] The BC Assembly of First Nations is a provincial-territorial organization whose membership is made up of 203 First Nations in British Columbia and is one of ten regional organizations affiliated with the national Assembly of First Nations, whose members include 633 First Nations across Canada.[3] All of these organizations interact with Clark's government.

There are two different aspects to Premier Christy Clark's relationship with First Nations in BC. The political one is obvious, but there is a less-known personal aspect as well.

Christy Clark's grandfather Johnny Clark was born in Clayoquot Sound, a remote area on Vancouver Island known for its beauty and wilderness. It is in the territory of the Nuu-chah-nulth First Nation, and this area is primarily populated by Aboriginal residents. There are fourteen so-called chiefly families in this region, which stretches about three hundred kilometres along the Island's west coast. They are the Ditidaht, Huu-ay-aht, Hupacasath, Tse-shaht, and Uchucklesaht in the south; the Ahousaht, Hesquiaht, Tla-o-qui-aht, Toquaht, and Yuu-cluth-aht in the centre; and the Ehattesaht, Kyuquot/Cheklesaht, Mowachat/Muchalaht, and Nuchatlaht in the north.[4]

Johnny and his two brothers, Charlie and Archie, grew up to become fishermen and spent decades working all over the coast in the commercial fishing industry. They travelled up and down the waters of BC, from the 1920s until the seventies or thereabouts. The reality of coastal life was that commercial fishermen worked side by side with First Nations people, Asians, and people of all backgrounds. They all lived in camps together, and there were no barriers to socializing and no concept of racial boundaries. The entire Clark family was accepting of people from all backgrounds, and this culture was passed down. In addition, Christy Clark's childhood in the Gulf Islands meant she had spent quite a bit of time with Aboriginal people.

"When I was a little girl," says Clark, "the woman who taught me how to swim was Rosemary Georgeson, from one of the biggest First Nations families on the island. She was just one example of the many people of First Nations descent who were just a part of our lives."

In October of 2015, the premier visited the Ahousaht First Nation in Tofino to thank them for their quick response to the sinking of the *Leviathan II* whale-watching boat and to call upon the federal government to consider integrating members of the community into the Coast Guard response team. She also explained that her father and sister had been rescued in a similar situation when Clark was only eight.

Although Christy Clark is perceived as a girl from the suburbs, she actually spent so much time in the rural and remote parts of the Gulf Islands that the experience helped shape her views.

"If you grow up exclusively in the city, versus regularly encountering First Nations in their communities, you have a different impression, or at least I did. I didn't have a preconceived notion of First Nations that some people might have. They were always a part of my summers as I was growing up."

Christy Clark has Aboriginal cousins, something she and her siblings found out only after she had become involved in politics. Her brother Bruce Clark remembers his great uncle Charlie, everyone's favourite uncle, as a bit of a character.

"One of our grandfather's brothers had children with a few women during his decades of life on the coast as a fisherman. I was too young to know the details, but I do know that some of them were Aboriginal. Our family did not keep in touch over the years in part due to distance, and the fact that this great-uncle moved around the coast quite a bit. We ended up with quite a few First Nations cousins."

Bruce said his family is proud of the connection and was happy when they met their relatives for the first time, even though it was a surprise, because they had no specific information about them. "Christy is very comfortable with Aboriginal people. She sees them all as family."

He laughed when he described a "typical Christy" moment that led to the family finding out about these cousins. "In the early nineties, Christy was attending one of the Liberal conventions, and it was shortly after the big breakthrough. She ended up talking to this fellow, and they spent a bit of time, and she invited him to join the group for the party at her place. After some more conversation where they compared relatives, they found out they were cousins."

Of course she was excited and wanted to expand her connections with this new-found cousin. "I will never forget the time she brought one of the cousins to a family funeral, where she introduced him to his older half-brother. The reaction on the face of the half-brother who had no idea of this connection was priceless."

Beyond these personal insights, there are several layers of issues to be considered to understand Premier Christy Clark's relationship with First Nations. The issue of First Nations in Canada is both simple and complicated at the same time. In the simplest terms, we can say they were here first, and what this means is that the Indigenous people were here before Europeans arrived, before there was a place called Canada, before there was a province called British Columbia. There were no immigrants because there was no country. Of course, it is much more complicated than that, especially in BC.

The history of Indigenous people in BC can be controversial, so this section draws on the chronology provided by Indigenous and Northern Affairs Canada, on the Government of Canada website.

First contact and permanent French and English colonial settlements began in eastern and central regions of what is now Canada in the sixteenth century, fuelled by aggressive competition over natural resources. There was trade in fish and furs with the Indigenous people, and alliances to expand the profitability of the trade.

Violent conflicts over the fur trade were resolved in 1701 in Montreal with the signing of the first legal agreement, a treaty that brought in the Great Peace. This ushered in the treaty process, with many more signed over the next 150 years. Treaties were the way the colonial populations entering the territories of the Indigenous people could have legal agreements to take resources and purchase land in exchange for recognizing the Indigenous peoples' pre-existing rights to hunt and fish and live. Indigenous peoples were never "conquered" by the colonists.

In 1763, after France ceded its colonial territories to Britain, King George III made the Royal Proclamation, which defined the western boundaries of the colonies, changed the relationship between Indigenous people and colonists, and created Indian Territories and the Indian Department. The Royal Proclamation publicly declared recognition of First Nations' rights and title and was intended to slow the western expansion of the colonies. Instead, the system eventually changed so that Indigenous people moved from being free people acting as allies of the colonists, to wards of the state on defined reserves of land, controlled by the Indian Department.

This situation evolved over decades, and eventually the treaty system, which once forged alliances, changed into a system of colonial dominance, and by the 1820s the goal was "civilizing the Indian" and imparting British values and culture.

In 1839, the Canadian government passed the *Crown Lands Protection Act,* which included Indian Reserve lands. Crown Lands are those lands that are not held privately, but are held by the government. Other legislation dealing with First Nations land and property was passed, and then in 1867 the *British North America Act* created the Dominion of Canada, with Crown Land authority extending from the east coast to the Rocky Mountains.

This leads us to the land between the Rocky Mountains and the Pacific Ocean, and the surrounding islands. In 1849, the colony of Vancouver Island was created, and the Hudson's Bay Company's Chief Factor, James Douglas (later the colony's first governor), signed fourteen treaties with the Coast Salish on Vancouver Island between 1850 and 1854. In 1859, the colony of British Columbia was created. This is where a major departure in policy and procedures occurred on Indigenous rights:

> Led by colonial surveyor and later lieutenant governor, Joseph Trutch, the colonial assembly slowly retracted the policies established by Douglas during the 1850s. *Treaty making did not continue after 1854 because of British Colombia's reluctance to recognize First Nations land rights, unlike all other British colonial jurisdictions.* This denial of Aboriginal land title persisted even after British Colombia joined Confederation and *ran contrary to the Dominion's recognition of this title in other parts of the country.*[5] [*Emphasis added.*]

In 1876, the Government of Canada introduced the *Indian Act,* a powerful and sweeping piece of legislation that aimed to consolidate previous regulations and gave greater authority to the Department of Indian Affairs. The *Indian Act* is in power to this day, although with more amendments than any other legislation in Canada, and affects people based on their race. It led to massive changes in First Nations' culture and rights, effectively forcing Canadian citizenship on them, imposing a system of band councils, putting an Indian agent in charge of everyone, and banning spiritual and religious ceremonies. In 1927, after the Nisga'a people in BC's north tried to pursue a land claim, the *Indian Act* was amended to prevent First Nations from pursuing land claims *of any kind*.

The obstruction and frustration of First Nations in BC obtaining legal agreement on their land continued almost to the end of the century, when the government of Premier Bill Vander Zalm agreed to sit at the treaty table with the Nisga'a people in 1990.[6]

"The history is that the province ignored First Nations for 120-some years," says associate professor Gordon Christie, director of the indigenous legal studies program at the University of B.C.

In 1869, B.C. unilaterally denied existence of aboriginal title, claiming aboriginal people were too primitive to understand the concept of land ownership, notes the BC Treaty Commission.

"Back in the 1880s, the Nisga'a used to take their canoes down to Victoria. That's quite a trip. It shows their determination."[7]

After the *Indian Act* was amended in 1951, First Nations were able to become lawyers and file lawsuits, and the Nisga'a won a decision, in 1973, that forced the provincial government to take notice. Over time, the courts began to rule on the inalienable rights of Indigenous people. When First Nations filed their claims in court, they invariably moved their claims forward, and each win created a new reference for the provincial and federal governments that they could not ignore.

The steady advance of the Nisga'a on their land claims in court, and the provincial government's agreement to sit and talk to them, led to the negotiation of the first modern treaty, which became a hot topic in the

1996 election. Gordon Campbell's BC Liberal Opposition was strongly opposed to the treaty. Premier Glen Clark's NDP government wanted to pass it. Also active in the election's leaders' debates were Reform Party Leader Jack Weisgerber, strongly opposed to the treaty, and PDA Leader Gordon Wilson, strongly in favour. Jack Weisgerber had been Premier Bill Vander Zalm's Minister of Indian Affairs (as it was then called) when the treaty talks started.

When Glen Clark surprised the pundits and won the 1996 provincial election, he had a mandate to sign the treaty. The BC Liberals were adamant that its legislation would not be passed in the legislature, and in July 1998, Campbell declared his party's intent to sue in court to declare the treaty unconstitutional and demand a referendum before the legislature voted on it.[8] To complicate issues, a vote held in November in the Nass Valley on the treaty had only 51 percent support from the Nisga'a people.[9]

Still, the Glen Clark government was determined to move ahead, and Nisga'a President Joe Gosnell was working hard in the Nass Valley to see the 111-year-old dream of land rights come true. When the NDP convened the spring sitting of the legislature, they had a new cabinet minister. Gordon Wilson joined at the end of January and became Minister of Aboriginal Affairs and Minister Responsible for BC Ferries. One of his tasks was to pass the Nisga'a Treaty. The spring session dropped the usual Speech from the Throne, and there was an urgency to pass the bill by the end of April.

Campbell's BC Liberals fought hard to delay passage of the bill. The lawsuit was controversial, the calls for a referendum picked up a lot of public support, and the story was front-page news almost daily, highly critical of Glen Clark and Gordon Wilson.

On April 13, 1999, Aboriginal Affairs Minister Gordon Wilson announced the government's plans to use a controversial act called "closure" to pass the treaty, which means that he set a specific end to the debate and a date for the vote. The BC Liberals erupted in fury, and the media coverage was very negative of the government's plans.

In a series of editorials, the *Vancouver Sun* declared "Right treaty, wrong process: using closure to make the Nisga'a treaty an irrevocable law is an abuse of democracy,"[9] and "government taints treaty by trampling democracy."[10] Meanwhile the front page of the *National Post* said, "NDP cuts off debate, passes Nisga'a treaty; ruling NDP shuts down debate on Nisga'a treaty: Liberals accuse embattled premier of despotism."[11]

When the day of the historic vote came on April 21, 1999, Nisga'a people flew down from the Nass Valley to attend. I was in the legislature that day, and I remember the buzz of excitement in the buildings. The Nisga'a treaty had been such a huge issue for so long, and yet many of us had never met a Nisga'a person other than the negotiators. Of the Nisga'a who came to celebrate the day of the vote, many said they never thought they would see the day when the treaty would pass in the legislature of a province that had fought against their rights for so long.

The BC Liberals walked out of the legislature when Question Period started, and returned for the vote on the treaty to vote against it.[12] Christy Clark was an MLA in Gordon Campbell's caucus. The vote passed with a vote of 38 to 32. Christy Clark voted against the treaty.

While the above provides some legal/constitutional background of the BC Liberals prior to Christy Clark's government, it is important to also dive into the social/cultural context before Clark became premier, before examining the dynamic between her provincial government and BC's First Nations in 2016.

In 1883, the Department of Indian Affairs began applying a new approach in dealing with Indigenous people; it created and then focused on the residential school system to "civilize" and "assimilate" First Nations children. The children were removed from their communities and families and often separated from their own siblings and put into residential schools, a form of boarding school where they were forced to abandon their language, culture, religion, and dress, and they were taught reading, writing, and either English or French. In these institutions, children were subject to harsh discipline, poor education, and lack of supervision, which left many vulnerable to, in many instances, physical or sexual abuse by the very staff hired to care and educate them.

Prime Minister John A. Macdonald, addressing the Canadian government in the House of Commons, said "When the school is on the reserve the child lives with its parents, who are savages; he is surrounded by savages, and though he may learn to read and write his habits, and training and mode of thought are Indian. He is simply a savage who can read and write. It has been strongly pressed on myself, as the head of the Department, that Indian children should be withdrawn as much as possible from the parental influence, and the only way to do that would be to put them in central training industrial schools where they will acquire the habits and modes of thought of white men."[13]

These schools operated until 1996, and approximately 150,000 First Nations children attended. For some in-depth stories about the experience, I recommend *Stolen from Our Embrace: The Abduction of First Nations Children and the Restoration of Aboriginal Communities*, by Suzanne Fournier and Ernie Crey (Douglas & McIntyre, 1998).

The devastating legacy of the residential school system is one that led to a lengthy and comprehensive federal initiative called the Truth and Reconciliation Commission of Canada, established in June of 2008. It had a mandate to learn the truth about residential schools and inform all Canadians about what happened.[14] In June of 2015, the Commission published a series of detailed and comprehensive stories about the damage caused and the policies implemented by the government, plus ninety-four specific calls to action to address ongoing problems originating from the residential school policy and its implementation.[15]

When Christy Clark was Minister of Education in the Campbell government, she first encountered one of the legacies of government interaction with First Nations. "I really started working with them when I was education minister. I learned that our system was failing them, especially the graduation rates. It was terrible."

This affected her, and she wanted to explore the issue to find a solution. "I did a lot of travel. I was at every school district in the province while I was minister, and that would almost always include speaking to First Nations." Wherever she went, she heard the same story. The system was not working for Indigenous students.

"First Nations were adamant that, unless we measured success, we would never have any accountability, and there would never be an improvement; therefore, we implemented accountability contracts with every school district." The contracts were not general in the outcomes they measured. "We also came up with really clear accountability measures for school districts for graduation rates for First Nations."

When she talks about the issue, her body language changes and her voice is very passionate. "I encountered all of these really smart First Nations leaders who were really working hard to try to solve the problem. Some school districts were shining stars and some were in real trouble."

Clark implemented the solution to ensure it would meet the needs expressed by the local leaders. "We wanted to make sure we could measure the progress, so that students would have better outcomes."

Did Christy Clark's changes work?

According to the Canadian Education Association's article in *Education Canada* entitled "Is BC Getting It Right? Moving toward Aboriginal Education Success in British Columbia":

> British Columbia's 2012–2013 six-year Aboriginal student high school graduation rate is 60 percent, a marked increase compared to 2008–2009 when it was 40 percent. In the last three years, the increase has been three percent per year.[16]

Did the accountability contracts measuring outcomes make the difference?

In May 2014, SFU economics professor John Richards, in his report "Are We Making Progress? New Evidence on Aboriginal Education Outcomes in Provincial and Reserve Schools," wrote that "B.C. outperforms all other provinces because of additional funding, better student tracking and working with Aboriginal leaders."[17]

In "Aboriginal High School Graduation Rate at Record High in B.C.," Geordon Omand of the Canadian Press reported on December 29, 2015, "The number of aboriginal students finishing secondary school in the province has increased steadily from about 54 to 63 per cent over the past six years."[18] This report does not comment on the change in public policy that led to this improvement.

In BC, there is now a First Nations Education Steering Committee and a First Nations Schools Association that provides direct input to the Ministry of Education on policy and initiatives to assist with educational success. The measures Clark implemented as Minister of Education remain in effect in 2016, with enhanced focus on First Nations culture and curriculum, resulting from the ongoing input of Aboriginal community leaders.

Christy Clark also served as Minister of Children and Families in Campbell's government, and her emotions are close to the surface when she discusses how Indigenous people taught her about the impact on their villages of historic government actions.

"Sitting with First Nations leaders and hearing from them the impact of child apprehensions from their communities really had an impact on me. It was First Nations saying our kids need to be connected to their history if they want to be whole people. This really changed my view because the sense of *place* was so ingrained."

Clark spent considerable time speaking with the matriarchs in the First Nations communities to try to understand their stories and culture in detail so she could change public policy in the best interests of the children. She knew she needed to balance her mandate as minister with the communities' needs. This balance is one she works to maintain as premier, where safety is first, and security includes the community.

"Our first job has to be to keep children safe. And we always have to strive to find a way, if we can, to keep the children safe in their own community. If we want them to be healthy people, they have to be whole people. For these communities to succeed, they can't lose their children. Their children are their community's future."

Through hours of travel and listening, in community after community, she gained a detailed vision of First Nations. "I learned through this experience to appreciate the incredible sense of intergenerational connection that First Nations have to their children and to their land; for their children to be removed from their community and the land that is part of them is *uniquely* traumatizing."

As she explains this, the depth of the emotional impact on her is obvious. Each community leader, whether matriarchs, hereditary leaders, or elected leaders, talked about the importance of community and the need to develop ways to keep their young people at home. They talked about the high suicide rates and crippling unemployment. The social issues were directly connected to unemployment and hopelessness.

"Since becoming premier, I started to talk to First Nations about economic issues."

With economic issues comes the discussion of LNG, resource development, and treaty negotiations, which are at the heart of the controversy with the Christy Clark government. Before diving into this issue from Clark's view, it is important to hear from a prominent First Nations leader whose story encompasses the social and lands title issue, Grand Chief Stewart Phillip. Grand Chief Stewart Phillip is president of the Union of BC Indian Chiefs (UBCIC), a position he has held since 1998.

WHEN I SPOKE with Grand Chief Stewart Phillip, he had just returned from Prince George, where he attended a meeting preparing for the inquiry into Missing and Murdered Indigenous Women. BC's Aboriginal Relations Minister John Rustad had hosted the meeting in anticipation of the new federal Liberal government's planned inquiry, and about five hundred attendees had shared their stories.[19]

We talked about the impact of social issues. "This is an intergenerational problem," says Phillip, "and the underlying issue is trauma. We heard from Gabor Maté, who did a good job... He talked directly to the attendees, and asked people to describe an episode where you were really angry with a loved one, a partner, a co-worker, and then he traced it back for them. There was one lady, she said 'they took my dignity' because they cut her hair off when they took her, and she felt it so strongly. So he traced that back for her, and at the end she said 'all they did was cut my hair.' So many of the people in the room have triggers for their bad experiences, and as one person talks, it is a trigger for someone else's memory, and then it triggers everyone in the room. And so there is a lot to discuss."

Trauma is not just a theoretical idea for Phillip, since he has his own story to share, and it has impacted so many of his choices and actions later in life. I was honoured to hear his story, and provide some of it here, as it is a story that is similar to the experience of so many Aboriginal people. Anything in brackets is provided by me for clarity, and has been cleared by Grand Chief Phillip:

I was raised in care, apprehended from our home in Penticton when I was one year old, in 1950. At that time, tuberculosis was rampant in the First Nations communities across BC. There was a sanitarium in Sardis, and lots of First Nations were quarantined there, including my parents. My nine-year-old sister was trying to take care of me, but there were no services, no band offices back then. We lived in an area defined as the Kootenay–Okanagan District, and the office was in Vernon [about 115 km away]. My sister struggled to look after me, and the authorities found out, and they came in and apprehended me.

The lady who took me in, she was my mom, a wonderful lady. When they handed me over, they told her there was a good chance I wouldn't make it, and they wouldn't hold her responsible. She used to tell me stories of getting up every two hours through the night to feed me, to help me gain strength.

I had nine brothers and sisters. My older siblings all went to residential school, and me and all the younger ones were apprehended.

My stepdad worked at a small sawmill at Hedley until it closed down, and then we moved to Quesnel in the Cariboo, and he got employment at the mill there...

It was in high school that I started to feel the effects of racism. There were about a thousand students, and maybe three of us were native. I was fairly popular, but when it came to asking girls out, that's when I felt the gap. I was different because I was Indian.

In Grade 8, I met a girl who was sitting at the back of the room. We became friends, then we started going steady. She was my first real girlfriend, and after high school we married. But I started drinking at fifteen, and that was really destructive.

My wife was not native. We had two beautiful girls, who are both teachers today. I got a job in the mill there (in Quesnel).

Because of my drinking, I was totally messed up. I had a lot of issues, especially around my loss of identity. I didn't know who I was.

I will never forget the day (my wife) called me at work and said, "You better come home. Your dad is here."

When she said it, I was worried because my mom and dad, they were always together, and I thought if he's there on his own, something must have happened.

Then she said, "It is your *real* dad."

I worked about forty or fifty miles from home, and my mind was full of thoughts on the drive back. Quesnel is a pretty racist place against natives, and I had this image of a drunk man bashing about in the house, dressed like a bum, and I was afraid for my family. At the same time, I had this huge sense of pride, for the first time. I had a real father! I had always been teased at school. "Why don't you have any real parents?"

It was amazing how powerful the feeling was. I was ecstatic and then fearful because of my learned attitude of racism.

I expected an old jalopy in the driveway, a beat up vehicle, but instead there was a robin's-egg blue Ford Fairlane, a beautiful car, with a nice sports jacket draped over the seat. I went inside and my father was great. We spent about six hours just talking, and I was so impressed.

A few days later, we all went to Penticton, and for three days non-stop I met a series of relatives. I was in my early twenties, and elders were all coming in, talking to my dad in our language. It was a momentous occasion.

It was the first time I was treated with respect for who I am, as if I had value. I belonged.

It turns out I was the last one to come home, the last of our people to be found. My dad had set out to find me.

After I was divorced, I moved back to Vancouver. There were a lot of bad things happening with our people at that time, though. Mayor Tom Campbell was in, we called him Terrible Tom, and he was shutting down the shooting galleries. What that meant was, if someone was hooked on hard drugs, they had to go somewhere else. Some bad influences were near where I was living, so I wanted to leave.

I had nowhere to go, so I moved to Penticton and lived with my dad in our community. While I lived with him, I learned all the stories of our people.

My biological mom lived in the same community, in the house furthest away from my dad's. One day he took me for a drive, and I was in the back seat. He drove all the way to the other side of the community, and he stopped outside a small house. My mom was timid, and some of my sisters were peeking through the window, and they were timid, and after a while my mom came out to the car. My dad was sitting in the driver's seat, just staring straight ahead, so my mom said, "What's going on?"

My dad replied, "I brought Stewart."

My mom looked in the back seat and just started to cry. She was the sweetest lady.

Soon after, I applied for a job with the band office and was hired as the education counsellor. Then I ran for council and was elected for ten years, starting in 1973, then I was chief for fourteen years, and President of the Union of BC Indian Chiefs. I have been with the UBCIC for eighteen years now.

After this incredible journey, it is little wonder that Grand Chief Stewart Phillip is highly motivated on issues for First Nations on Indigenous rights and children in care. His active role in politics, often with the NDP Opposition, has meant constant engagement with all levels of government.

Phillip says, "I first met Christy Clark when she was Minister of Children and Families, in 2005. I remember her as a bubbly, gregarious person with a big smile. She's pretty quick on the uptake, and is articulate and a good communicator. She certainly knows how to say the right thing, but for me, she's never followed through."

He cites an incident that, for him, defines his engagement with Premier Clark.

"When she was Minister of Children and Families, I met with her on behalf of the UBCIC because we wanted the government to know that we were opposed to them adopting out our children. The government had stated that they wanted to reduce the children in care by adopting them out, and we told her we wanted to bring our children home. I thought she was supporting our position."

Phillip was in Victoria for a meeting with NDP MLA Joy MacPhail after his meeting with Clark, travelling with his office administrator Don Bain.

"It's funny how life turns out sometimes... We had some time before our [second] meeting, and didn't know what to do, so Don suggested we go up into the legislative gallery and watch the discussions. There was some kind of committee meeting going on, and Christy Clark was speaking as Minister of Children and Families. She was going on about how successful they were at adopting children out and reducing the children in care. We had just had a meeting where we told her we did not want our children adopted out, and here she was reporting that as a success."

WENDY JOHN, formerly Wendy Grant, has been immersed in First Nations issues and advocacy from a young age. She is a member of the Sparrow family, served as Chief of the Musqueam First Nation for three terms and was the first woman elected as regional chief for the Assembly of First Nations. The Sparrow family led a court case that defined First Nations rights in 1990 and led to a test of those rights called the Sparrow Test, which was largely seen to protect the inherent rights of First Nation people to the food fishery and other similar cases of resource access.

Wendy John is a dynamic and youthful woman who is strikingly attractive. She is sixty-seven, and still makes snow angels with her husband. John speaks quickly, confidently, and articulately, and can cover immense and complicated topics in rapid succession, with such gentle treatment that the listener may not realize that the information covers high-level constitutional and other complicated topics. She's also very funny.

John has a unique view of Christy Clark's commitment to children in care because her husband, Grand Chief Edward John, was in the NDP cabinet as the Minister of Children and Families and now serves the BC Liberal government in the position of senior advisor on Aboriginal child welfare to the Minister of Children and Family Development.[20]

From the government news release for Grand Chief Edward John's appointment:

> The Ministry of Children and Family Development (MCFD) has prioritized adoptions and other forms of permanency for children who are in long-term care—particularly Aboriginal children, given their over-representation within the child welfare system. The role of the senior advisor will be to assist in developing stronger permanency plans that will lead to greater success finding forever families through adoption, guardianship or other options that may be more appropriate in First Nations communities.

Wendy John became involved in politics after pressure from the women in her community. "Because of my past work with Coast Salish weaving, they really pushed me. They started to push me when I was thirty-six, and I was elected when I was thirty-eight. We have a matrilineal society, and my involvement in politics was all because of the women."

At the time that she became chief, women were still rare in elected leadership roles, and her perspective was essential in defining a different focus. "Before politics, my work was in service to the community. My efforts are all driven by family and community because we First Nations are a collectivity. In our culture, women are equal. We are all part of the whole, the collective."

John takes time to explain the integrated nature of First Nations' settlements. "Everything I've ever done, and every fight I've ever been part of has been driven by the best interests of the community based on the family and that collectivity of the family in the village. We don't separate out the individuals, or the individual families; we are one."

She sees Christy Clark's leadership through a lens that is about family. "When I think of Christy Clark, I think about her focus, which is 'families first,' and that is right; that needs to be the focus, and the centre of decisions. When Ed was Minister of Children and Families for the NDP, I saw how difficult the issues were and how hard it was for him to manage that file. Christy Clark was very critical of the government at that time on this file.

"Christy Clark became the Minister of Children and Families afterward, when the Liberals became government, and I thought that maybe this is why, when she was leader, she wanted to put families first—because she knew that's where the problems are most important."

Wendy John, herself a mother, is passionate on this topic. "If you don't have healthy families, nothing else matters. We have to strengthen families and make sure the care of the children is the most important part of that."

Public interest is all about families, according to John. "Why create a great economy? Why invest in universities? Why set up a great education system? Why bother with social programs? All of this must be with a goal to supporting our families and they need to be strong.

"How do we take care of our families? Through jobs, *good* jobs. That has been Christy Clark's goal, and she seems to understand that this is the important base for the families. Parents need to be able to contribute to their families, which helps build and support the whole community."

As an experienced political leader, Wendy John understands that sometimes there can be unanticipated challenges. "I think she's been struggling lately. Her vision for the jobs has taken up so much of her government's agenda, and the collapse of oil prices has played a key part in her not being able to expand her focus past jobs to some of the other issues dealing with families. She knows the protection of jobs has to come first so families have that covered. This is how I see her priorities, and that makes sense."

Not everyone in the Aboriginal community agrees that jobs are the key to family security as a priority. Issues such as economic development, resource use, land title, and Indigenous rights all recur as a theme at the heart of the relationship between First Nations and Christy Clark's government.

In 2013, Premier Christy Clark campaigned that LNG as a new industry would provide economic development, jobs, tax revenue, and partnership opportunities. One of the key players in developing the LNG industry are BC's First Nations. Jas Johal left Global BC TV news to take a position as director of communication with the BC LNG Alliance, an industry association for the emerging energy sector. He travels the province working on LNG development. "LNG development in BC is going to move First Nations communities forward in a major way."

Wildlife consultant Guy Monty also travels the province on resource project assessments, and his view differs from Johal's. "The upcoming conflict with First Nations on LNG is going to be extremely disruptive to the provincial economy. I have never seen Natives as engaged and angry as they are about the pipelines and resource development planned for that region."

KIM BAIRD WAS the elected chief of the Tsawwassen First Nation (TFN) for six terms from 1999–2012. She is a community leader on Aboriginal issues pertaining to reconciliation, treaties, economic development, and social justice. She implemented British Columbia's first urban modern treaty on April 3, 2009, and has since overseen numerous economic and institutional development projects for TFN.

"Socioeconomic conditions are the most pressing issue," says Baird, "though the roots of it are based on systemic issues such as the non-resolution of Aboriginal rights and title in British Columbia (treaties or other reconciliation models)."

She also cites the difficulties of First Nations being covered under the *Indian Act* system, with a structure that prescribes "inadequate governance" and underfunding.

Politics, suggests Baird, sometimes ends up mixed in with First Nations service delivery, and "things are depoliticized when we are engaging in ceremonies and community celebrations such as weddings and grad dinners or things that are hosted by individuals and not government."

Baird has met Premier Christy Clark a few times, and one memory stands out for her. "One of the most helpful things that ever happened to me was when I lost my election as chief and she called me in person to commend me on my service to TFN and BC. That was pretty helpful for me to find my feet again in what was a very shocking upstart situation."

Have there been tangible results for First Nations with Christy Clark's government? "I think she is supportive of First Nation issues," says Baird, "more from an economic lens than other ones perhaps. There are many economic agreements.

"I had the occasion to ask her about her approach with First Nations because it seems kind of an organic approach where BC makes progress with First Nations where it can. I had told her that to others it might not appear to be a strategy at all. She thought that BC was doing a fantastic job and come light years ahead of where things were when she was a minister."

Baird acknowledges that many First Nations people are angry with Premier Clark. "I think there is some degree of complacency. I also think that her economic lens is not getting to the crux of resolving rights and title, and I think this is what needs to happen over the long-term in some way with over two hundred First Nations. Daunting, I know. The way the BC Treaty Commission issue was handled is an example of a seeming lack of importance to this issue."

She is referring to the controversial and surprise decision by Clark's cabinet in March 2015 to veto George Abbott's appointment as chief commissioner of the BC Treaty Commission. In Victoria's *Times Colonist*, Lindsay Kines reports:

Clark said cabinet decided to stop investing in a system that has cost $600 million and produced just four treaties in 22 years. "We have to be able to move faster and we have to find a way to include more First Nations in the process," she said. "Fifty out of 200 First Nations involved in the process?... In terms of next steps—whether or not the treaty commission will be changed, whether or not it will continue to exist, how all that future will unfold with respect to treaties—[it] is going to be something that we do together with First Nations."

Clark said the decision evolved over many months, and she admitted making a "mess" of communicating the changes to Abbott.

NDP Leader John Horgan said Clark did more than make a mess of communications. He said the decision not to appoint Abbott blindsided First Nations as well as the federal government and undercut more than two decades of treaty work in B.C.

"In my own community, three First Nations are at the table, near final agreement," said Horgan, who represents Juan de Fuca. "They now have no confidence that they're going to be able to achieve that after investing significant resources.

The First Nations Summit said it was "taken aback and seriously disappointed" by the decision to withdraw Abbott's appointment and ignore a recent chiefs' resolution that formalized his appointment.

Grand Chief Edward John said it was "not very gracious" treatment by government and raised doubts about its trustworthiness. "I'm sure it says a lot about government commitment, and agreements that are reached, that they simply can just disregard them," he said.

Clark insists there is "a broad consensus" among First Nations that the current treaty process has not worked. "If we keep going at the pace we're going over the last 22 years, it will be a century before we conclude all the treaties in British Columbia."

She said First Nations have sat outside economic development for too long and asking them to wait another 100 years is too much.

"We have to do better for First Nations and that's what we're trying to do."[21]

What many people may not know is that, in 2013, Abbott began a Ph.D. thesis at the University of Victoria that could have created some issues for the BC government. Abbott announced it would be published in early 2017, just before the next election and describes it as "a 200-plus-page thesis on how the Tsay Keh Dene had their lands flooded out due to the construction of the W.A.C. Bennett dam in the 1960s and the impact of Canada's division of powers between Ottawa and the provinces on Aboriginal public policy."[22]

Given Christy Clark's government's decision to proceed with the Site C project, and the ongoing issues with some Aboriginal people, having the chief commissioner for the BC Treaty Commission, writing about a similar historic issue could be very problematic. If we add the dynamic of BC's antagonistic politics, it could blow up on the Clark government in a significant way.

George Abbott was elected in 1996 under Gordon Campbell, ran against Christy Clark to replace Campbell, did not run for re-election in 2013, and quit the BC Liberal Party a few months after the cabinet vetoed his appointment as chief commissioner.[23]

Kim Baird thinks treaties and land title are key issues and refers to the Tsilhqot'in, or William Decision, of June 26, 2014. "The escalation of court cases such as the William Decision shows that, while political statements may be made about recognizing aboriginal rights and title, this is not demonstrated... [by] creating and respecting legal rights for First Nations. Site C and other projects are proceeding in spite of opposition. And there does not seem to be strategic importance about Aboriginal Rights and Title... [and] some interpret this as opposition to First Nation issues... in absence of a common vision of a road map, I think the animosity will continue to build in my opinion."

Tsilhqot'in is the latest in a series of Supreme Court of Canada rulings on Indigenous rights, following the Calder Case of Frank Calder of the Nisga'a Nation in 1973, the Sparrow Ruling of May 31, 1990, and the landmark Delgamuukw case against the BC government of December 11, 1997. This Tsilhqot'in Decision (Tsilhqot'in Nation v. British Columbia: 2014 SCC 44) has been referenced when environmentalists and First Nations oppose resource decisions in BC's rural regions, sometimes with claims that it recognizes Aboriginal sovereign title. Robin Unger, Joan Young, and Brent Ryan of McMillan LLP, a Canadian law firm with offices in Canada and other countries, wrote a summary in the McMillan *Aboriginal Law Bulletin* of June 2014:

> The Court has confirmed that Aboriginal title can exist over relatively broad areas of land that were subject to occupation at the time sovereignty was asserted.
>
> With the exception of clarifying what is required to establish occupation, the decision does not make significant changes to the law of Aboriginal title as it has come to exist over the last several decades.
>
> The decision makes clear that provincial laws apply on lands for which Aboriginal title is claimed or proven.
>
> In keeping with well-established law, federal and provincial governments continue to have a duty to consult and potentially accommodate in cases where Aboriginal title is asserted but not yet proven.
>
> Governments can infringe proven Aboriginal title, provided they meet the established tests for "justification."
>
> And while the area of land over which title was found is not insignificant, it is also important to note that it represents only approximately 2% of the Tsilhqot'in traditional territory.
>
> The Tsilhqot'in decision is historic and ground breaking in the sense that it is the first time Aboriginal title has been declared under a framework that has been in existence for decades. But in many respects the decision simply adopts and applies existing jurisprudence and does not represent a substantial change in the law of Aboriginal title. It does however provide clarification on what constitutes "occupation" for title purposes, as well as confirmation that provincial laws continue to apply to Aboriginal title lands, subject to justification requirements.

Such clarity is essential to promote reconciliation efforts and the continued governance of Canada and British Columbia.[24]

JEAN TEILLET IS a Métis lawyer from Red River, Manitoba, who is a member of the Riel family. She is a fast-talking powerhouse, dresses like a fashionable artist rather than an intense lawyer, and talks with her hands, smiling as she describes complicated historical issues. Teillet has worked in Vancouver since the early nineties as a treaty negotiator and Aboriginal rights litigator, for both Métis and First Nations.

"This is a very ancient role for Métis to play. We have long been the negotiators, translators or facilitators between First Nations and government... There are many Métis like me who live in BC, but we have no claim to rights here."

After her decades of work on these issues, Teillet says, "I think the most important provincial issue facing First Nations is inclusion/recognition. Right now BC governs based on three presumptions. One, that they own all the lands and resources in the province; two, that they have exclusive decision-making about the lands and resources in the province; and three, that they are entitled to all the benefits from the lands and resources in the province.

"Tsilhqot'in was a small nomadic group over a large territory. They were the hard case to prove Aboriginal title. My assessment is that if Tsilhqot'in can prove Aboriginal title, then so can ninety percent if not a hundred percent of the First Nations in BC. Right now, BC is proceeding on the basis of a risk analysis—that BC First Nations will not get it together to take the province to court because they either cannot afford it or can't get it together (or keep it together) politically for the length of time it would take to litigate. This is a very short-sighted policy. It's also full of risk because the Supreme Court in Tsilhqot'in said that the province can proceed to make decisions about Aboriginal title land before it is affirmed by the courts, but if there was no First Nation consent, the province will be vulnerable to claims for compensation from the First Nation."

How does this situation impact investment in BC and the economy? "BC's current attitude also contributes to the sense that nothing can get done in BC because aboriginal people will block every project."

Teillet, like many others dealing with land issues in BC, says that at the base of this is the failure of the BC Treaty Commission. This perception is one shared by Premier Christy Clark, and Teillet says, regardless of the problems, there has to be a way for overall progress.

"BC has lately seemed to be abandoning treaties and putting its energy into one-off agreements. While these one-offs will solve the immediate problem and allow one project to go ahead, they mean that the province is in perpetual negotiations with First Nations for absolutely every decision. *No one* has the time and resources to do this. It is *not* a plan. It is not forward thinking."

According to Teillet, the basic assumptions must be revisited. "[But] I would say that BC will strongly resist any attempt to revisit this. Despite the fact that the new federal Liberal government is proposing just that when it says that it intends to implement UN Declaration on the Rights of Indigenous Peoples ... BC must come to grips with the fact that 'consultation' is not the end of the game, it was just the process the Supreme Court put in place to try to nudge governments into action on aboriginal title and rights. The Court can be expected to go much further if governments do not implement fundamental change."

Teillet does not believe it will be easy to move forward. "No one in power likes to give up power. But this is going to happen one way or another."

Have there been tangible results with Premier Christy Clark's approach to Aboriginal people? "She seems to be committed to certain projects," says Teillet, "and she puts her considerable charm and effort into smoothing bridges between a First Nation and a proponent and the government so that the project can proceed. But she has put *no* effort into long-term relationships or improvements. Far from seeing tangible results, we have seen negatives ... the failure to approve a Chief Commissioner of the BC Treaty Commission for the past year; the failure of

BC to engage in the Principals meetings; and removal of even basic staff from MARR [Ministry of Aboriginal Relations and Reconciliation] which has created a four-year log-jam in the treaty process."

Teillet says that the First Nations who are angry with the premier "know that she sees them only as impediments, trouble-makers, 'the forces of no.'"

I asked Teillet if she could send a message to Premier Clark about First Nations issues, what would it be?

"Get a plan! Start thinking long term about this issue. Lay groundwork. Get some good people to think this through. Staff the problem . . . It will get worse if she doesn't apply herself to it. Christy should step up, roll up her sleeves and go to it. It can be done, and the rest of the people in BC can be educated to see that it's in everyone's best interest."

ONE OF THE most vocal opponents of Premier Christy Clark's government's policies with First Nations is Grand Chief Stewart Phillip, who believes the BC government took a step backward from the Leadership Accord signed with the government of Premier Gordon Campbell.

"On March 17, 2005, a historic agreement was signed between the Union of BC Indian Chiefs, the First Nations Summit, and the BC Assembly of First Nations. There were three provisions of this Leadership Accord: first, we had to acknowledge and respect the unique and separate mandates of each organization; second, we would collaborate and work together on issues of mutual and common concern; and third, if we had irreconcilable differences, we would part company. In March 2016, this accord will be eleven years old."

Because of the political controversy under Campbell's leadership, Phillip says, "When Premier Christy Clark assumed the leadership, I am convinced her advisors cautioned her not to engage with the Leadership Accord in any meaningful or substantial way."

When asked if there is a way to have a constructive working relationship with the Clark government, Phillip is emphatic. "No. We had our first historic meeting with Premier Christy Clark and her government a couple of years ago, after the Tsilhqot'in Decision. It was on September 11, and we called it the 9/11 meeting. All the chiefs in the province attended, plus a legion of legal advisors."

At the heart of Grand Chief Phillip's anger at the Clark government is the disagreement over the Four Principles as set out by the parties to the Leadership Accord. In a long letter from Deputy Attorney General Richard Fyfe, the province rejected the underlying assumptions. Vaughn Palmer reported on this letter:

> "The overarching problem," explained Fyfe, "is that the statements are founded on the assumption that there is Aboriginal title throughout B.C.—a term that may be interchangeable with 'everywhere.'"
>
> "That interpretation is not consistent with existing legal authority or the available ethnographic and historical record," he continued. "It would be inappropriate in our view for the province from a legal perspective to proceed with negotiation of agreements based on those concepts."[25]

Grand Chief Phillip says that Fyfe's position is obstructing decision-making on the basis of the legal precedent. It all came to a head at a meeting of the All Chiefs in the fall of 2014 between the First Nations leadership and the government of BC.

"There were Four Principles, and in the first meeting, Ed John and Jody Wilson-Raybould went into exhaustive detail about unextinguished Indigenous rights and title, and all the legal precedents and court rulings. The meeting took place during the teachers' strike, and I spoke third, after Ed and Jody. My two daughters are teachers, so after recognizing the territory of the Coast Salish, I said I want to give a shout out to all the hard working teachers in BC. I was close enough to the premier that I could have touched her. Her face looked like it turned to stone, and she wasn't too amused. The place erupted in a cheer, there was a lot of support for the teachers. Then I said, I'm not a lawyer, but BC is Indian land, and there were huge cheers.

"That afternoon, we announced we had no agreement. Last September was the second meeting, which was round two. The next meeting [2016] is critical."

As for where Phillip fits in BC's antagonistic political arena, he is very clear when it comes to the government of Premier Christy Clark—he wants the Liberals replaced by the NDP.

"I'm on record as saying we are halfway there. We dumped the regressive Harper government. Now we need to find a willing partner for our provincial government, and it isn't the Liberals. They have had their chance."

A MORE MODERATE view is held by Chief Robert Louie, of the Westbank First Nation.

"I was elected to council in 1974 and was one of the younger councillors in Westbank at that time. It was during the third term of Prime Minister Pierre Trudeau, and Dave Barrett was the NDP premier of BC. I became Chief of Westbank First Nation in 1986 and served until 1996. Then I was defeated in 1996, stayed out for two terms, and have now been chief since 2002."

Robert Louie is a leader who inspires confidence by his quiet authority; he does not raise his voice or make loud speeches. He works quietly behind the scenes, is consistently highly effective, and accomplishes considerable projects on behalf of his people. He is also notably handsome.

I have known Robert Louie since my years in politics in Kelowna, and he has always been well respected and an easy person with whom to work. He has considerable experience in Indigenous self-government and economic development. Westbank is one of the wealthiest First Nations communities in BC, possibly in Canada, and has had self-government since 2005, after the Government of Canada passed an Act delegating governing authority to the council instead of the *Indian Act*.[26]

"Westbank First Nation has a bilateral agreement with Canada, with a transfer of power to Westbank First Nation. On April 1, [2016] we will have completed eleven years of self-government. I was Chief when this was implemented and was involved at every stage. Our self-government occurred with the passage of the federal act, which was the *Westbank First Nation Government Act*. This was the first one recognized under the policy for inherent self-government in Canada. The 'Inherent Rights' policy existed, and there were about 57 or 60 First Nations talking about it at the time, but Westbank was the first one passed and the only one initiated at that time. There are some First Nations in discussions now; I think Whitecap Dakota First Nation Chief Darcy Bear in Saskatchewan is one."

It was a good move. "Self-government has really worked well for us."

When asked what the most important issue is facing First Nations, Louie is definitive. "Land claims are the most important issue facing BC First Nations. This was the most important issue when I was first involved in 1974. It has been the biggest issue all along and remains unresolved."

This may be a surprising comment coming from the leader of a government that has had ongoing success in housing, commercial development, and negotiation of partnership agreements.

Louie says, "Land claims are extremely important. While we have reserve lands on both sides of Okanagan Lake, we have unresolved claims, long-standing claims, on our traditional territory, which is land we have inhabited since time immemorial, and we have no agreements on revenue sharing or resource management on our territories."

He says there is no long-term certainty for First Nations. "We have the right to hunt and fish, from another court case, but that is only part of the issue. The claims on governing authority on our traditional territory have never been addressed."

Robert Louie was asked by Premier Bill Vander Zalm to be one of three Indigenous people to sit with him and his new Minister of Aboriginal Affairs, plus a committee of developers, to discuss First Nations issues. "The intent was to find ways to deal with the significant issues, and the government wanted to focus on economic development and social issues.

My advice to them, at the first meeting, was deal with the land claims first because that is what you are going to hear. And we had meetings all over the province, and it was true, in a hundred percent of the First Nations communities we met, that land claims were the key issue."

This set the stage for later work by Harcourt's NDP government in 1991. "This process opened the door to the Nisga'a to make a land claims settlement agreement, but then the door was closed. When the NDP were elected, they started the BC Treaty Commission, and this allowed others to negotiate, but the problem was that the mandate could not satisfy the needs of the First Nations."

Louie, like Teillet and Phillip, is not impressed with the BC Treaty Commission. "We have seen only a little success. I think Tsawwassen just concluded, but not much... There are two hundred plus First Nations engaged in the BC Treaty process, with only a handful signed up through modern treaty-making. The balance have no settlements, and no mandate for settlement."

Louie agrees with Premier Christy Clark's analysis. "We all know the premier's view of this process, which is that it doesn't work. The BC Treaty Commission does not work. There is no Chief Commissioner for the process; the province hasn't appointed anyone."

When he says this, he does not raise his voice, he is simply stating a fact. He sounds frustrated.

"Regarding the Treaty Commission process, I was on the task group representing the chiefs for four years, and I was intimately involved with the treaty negotiations. This was about 1996 until 2000. I can tell you, we met many times and we never could come to any agreement on mandates and terms. We were too far apart. We could not resolve it.

"During my tenure, we had many sessions. It didn't matter."

After almost forty years of representing his people and working on certainty, all Chief Louie can say about the BC Treaty Commission is, "It is a failed process."

In the fall of 2015, Canadians elected a new government with a very different approach to First Nations issues. When Prime Minister Justin Trudeau appointed his cabinet, he made history by appointing BC

First Nations leader Jody Wilson-Raybould as Minister of Justice, and Inuit leader Hunter Tootoo as Minister of Fisheries, Oceans, and the Coast Guard.

Does this change the prospect for land claims? "Now," Louie says, "we have Prime Minister Justin Trudeau in federally with the change in government in the fall of 2015. We have to wait to see if anything changes, but I can't say I'm optimistic at this time.

"I had a meeting with Christy Clark right around the time she became premier, when Westbank First Nation backed away from the treaty table, saying that mandates were too far apart. Afterward, we started negotiations on a bilateral process, and the door was opening a bit, but then there was another election that took place, and there has been no movement since. Until the mandates change, there is no hope of any treaties being negotiated, in my opinion, other than what you can count on one hand.

"Premier Christy Clark asked for my advice about the province's relationship with First Nations in about September 2014, so I put together a lengthy letter. It represented the best advice I could provide, and I forwarded it directly to the premier's attention. But I haven't seen much implementation yet on any of the points I raised."

Premier Christy Clark is elected out of West Kelowna, and serves as the Member of the Legislative Assembly with the Westbank First Nation as a prominent part of her electoral district.

How is Christy Clark as an MLA? Chief Robert Louie says the relationship is good. "We have an understanding with Premier Clark, and we have a commitment from her to meet a number of times per year. I think we had tried to make it quarterly, but it has become based on her availability, which can be limited, and we understand that. We do meet regularly on significant issues."

MANY FIRST NATIONS focus on the success that Premier Christy Clark's government has initiated, and view the new relationship as a positive one. Former Musqueam Chief Wendy John commented about the respect for Indigenous culture and art, and the emphasis on sharing it.

"The premier went to meet Prime Minister Trudeau in Ottawa, and on the news you could see her wearing a beautiful First Nations shawl; it was very visible. Was it cultural expropriation? No way, it was a sign of respect for Aboriginal culture. It looked really great, and I can tell you, a few years ago, that would never have happened. You would not have seen a first minister wearing Indigenous art."

One of the most significant accomplishments in Premier Christy Clark's 2013 term was the announcement of the Great Bear Rainforest. Premier Clark announced the creation on February 1, 2016. As Justine Hunter reported in the *Globe and Mail*,

> The deal, which will be enshrined in legislation this spring, applies to a stretch of 6.4 million hectares of the coast from the north of Vancouver Island to the Alaska Panhandle. It promises to protect 85 per cent of the region's old-growth forests, with logging in the remaining 15 per cent subject to the most stringent commercial logging standards in North America.[27]

The agreement took twenty years to conclude, and accommodated the input of First Nations along the coast:

> Marilyn Slett, chief councillor of the Heiltsuk Tribal Council, said her community was drawn into the conflict early on, but their views were not always part of the debate. "That's the milestone—the collaboration," she said in an interview. "As Heiltsuk people, as Coastal Nations, we aligned around common values of protecting the land." But her people want jobs too. "Our nation clearly and consistently said, there must be balance."
>
> A critical part of the final agreement was the completion of government-to-government agreements, 26 in all, that B.C. signed with the resident First Nations on how the agreement will be managed.[28]

Wendy John said the creation of this land preserve was a historic change in engagement.

"What happened with the announcement of the Great Bear Rainforest is what First Nations have been fighting so hard for since First Contact. We want to be included in meaningful consultation, and in this case, we were."

What was most significant was the ability to bring together diverse interests to sign off on an agreement on resource management and sustainability.

"I hold up the First Nations leaders in a place of honour for creating this agreement because of their steadfastness and resilience in pursuing their goals," says John, an experienced negotiator herself.

"Everyone didn't get everything they wanted; everyone came away a little upset that they had to compromise. That to me is the sign of a good deal, where everyone had to give up a little to make the common ground."

She sees this agreement as a welcome departure from the stereotype of First Nations obstructing progress. "I sometimes say to people in industry, you know we don't get up every morning and ask, 'What can we protest today?' We aren't like that. Although I'm sure that's how it seems sometimes. We are only on the news when we are saying no. The truth is, we just want to be at the table, because we have something important to contribute. We share our understanding of the land, and of our peoples."

Another point not often made is that First Nations do not always appreciate the interaction with urban environmentalists who show up whenever there is a resource dispute. "I'm not such a friend of the environmentalists," she says, "because of how they have used us over the years.

"After a deal like this, you will hear some environmentalists screaming at those of us who have supported it that we are 'sell outs', but that doesn't help anyone, does it? It turns my stomach. One chief said he wanted to stop the whole process. That's not very realistic."

You can hear the frustration in her voice, and she notes that asking to be part of the discussion is one way of ensuring that First Nations, who have the historic knowledge of the land and its history, are heard. "We are looking for respect for what we can contribute. We have something valuable to bring to the table, so don't just put us out in front of the project and use us to stop it. Include us in the discussions so that we can help be part of the decision. It will be a better decision if we are included."

For John, cynicism can interfere with the relationship. "Premier Christy Clark has the right values and vision when it comes to First Nations . . . The big challenge between any government and First Nations is to build trust. It is too easy for us to feel 'here we go again' if a comment is reported that sounds like government is working against us. It is easy for us to become so caught up in a battle, which has been so negative for so long, that when it starts to be positive, they think it is just another way to get us to be quiet while they pursue their own agenda."

She sees work going on that others may not. "The bureaucrats in Clark's government see her vision, and they are working behind the scenes trying to follow up. It is a long journey."

What does she say to people who are angry with the premier? "When our people tell me they are angry with the government, I point out that Christy Clark is the first premier in history who has tasked her government with a full-day meeting with First Nations, where her *entire cabinet* attends. It has never been done before. That shows amazing political will, and a true commitment.

"I tell our people, every right comes with a responsibility. You have to sit down and talk about solutions. I gently remind people about the profound commitment it takes to put all your cabinet ministers in a room, for an entire day, knowing that you are going to take a lot of flack. The first time they did it, it was an exchange of ideas and the comments were pretty negative. The second time, it was more productive. She set it up so that the chiefs could meet and work with cabinet ministers.

"After these meetings, any meetings, there's the follow up, and that can take a long time." John has had the experience of being an elected leader and understands the process. "I've seen a lot of movement, from the premier's office on down, to shift how we do business together."

The interests of her people are central to Wendy John's view. "Musqueam is very strong at fighting for our rights and negotiating to a yes. And we have seen progress."

PERHAPS ONE OF the First Nations leaders who has worked most closely with Premier Clark is Chief Ellis Ross of the Haisla First Nation, whose people were approached with an LNG project before Christy Clark became premier but had a hard time following up until she took over government. Ross is soft spoken, articulate, and passionate.

"I was first nominated as councillor for the Haisla in 2003, when Gordon Campbell was premier. I became a full-time, paid councillor in 2004 and served as that until 2011, when I was elected Chief Councillor, which I am still today."

Most of the First Nations leaders I interviewed, when I asked them what they thought the most important issue facing First Nations was, answered "land claims." Chief Ross had a different answer. "There are way too many issues, it would be literally impossible to solve them all, and that is part of the problem.

"That is probably the number one reason so many First Nations can't get anywhere. There's really no focus. That's the problem I see with many First Nations leaders; they are chasing too many issues at the same time. Often, they end up chasing the latest hot topic instead of keeping their focus on one thing."

Ross has clarity on what he was elected to do. "The hot topic of the moment doesn't help me with my unemployment problem, my suicide problem, or what I have to deal with in my community. The local leaders get it, the ones that are on the ground dealing with the issues in their community. The leaders on the ground are dealing with issues that affect the individual members."

First Nations communities and their social and economic issues are Ross's priority, and he says changing the status quo is key. "If we really want change, we have to look to ourselves. We won't get change with a lot of political rhetoric, or a lot of great speeches."

He knows his perspective is controversial among the First Nations leadership. "I don't get along too well with the regional, provincial, or national leaders because of my perspective."

As the chief of the Haisla, Ross could participate in many of the larger meetings, but "I stopped going to the UBCIC, the AFN, the [First

Nations] Summit in about 2004 or 2005 because it was always the same topics, the same leaders, making the same speeches, and it wasn't helping my community fix its problems. I've seen some of the speeches lately, and they haven't changed. I'm not sure if this frustrates me, or saddens me."

Regardless of how he feels about the political objectives of the organizations representing First Nations, he is working locally. "What we are doing in the Haisla community has turned things around because of our focus. We looked internally to solve our problems, we didn't ask other people to solve them for us.

"We started by answering the question 'Who leads us?' We don't have a fight in our community between hereditary leaders or unelected leaders or councillors because we know it is the council, the council we elect."

Hereditary leadership in Aboriginal communities can be controversial because the *Indian Act* imposed new structures, and establishing who is in charge today is important.

"Many communities in BC are just a battleground for leadership, a three-way fight to represent, and they can't get anything done. They are going nowhere. The irony is that the membership is watching this fight, not knowing that they have the most say in who leads them and not the 'leaders' and those members will be the ones to suffer."

How does this impact the dynamic with the provincial or federal government? "Our communities are so busy fighting amongst themselves, and so busy arguing about all the problems, we don't even know what we want, so if the government and industry comes to negotiate, we don't know what to say."

Ross believes clarity is critical in knowing what you want before you sit down with other governments. "There's too much rhetoric, and that doesn't answer the question.

"Do you want money? Land? Jobs? You need to have the answer."

Under this vision, the Haisla's situation has improved, and the relationship between the membership and the council has changed. "The situation is better now for the Haisla, at the council level and for individuals, than it was when I was first elected. That's because we changed the way we perceive the world. Thirty years ago, all the leaders were right. It was 'the big bad white guy' holding us back. Today, government is sitting down with us, and it is ourselves, as First Nations, who are holding either ourselves or our fellow First Nations down. We are our own enemy now."

Ross shares a story to illustrate how the Haisla have changed in their expectations from council.

"Years ago, we started a gas station to provide some services ourselves. I remember that the manager we hired wanted to sell his art in the gas station to supplement his income, and we ended up debating, at council, whether or not he should be allowed to sell his art at the gas station. We spent a couple of hours of our time on this, when we had huge resource deals to discuss. We don't talk about things like that anymore."

Instead, they have focused on job creation and resource development. "We went from 60 percent unemployment, and through our mini-boom, at one point, we were down around 10 percent unemployment, and that was, likely, unemployables. Even today, after the mini-boom, our recreation centre parking lot is full of brand new trucks, all bought with money earned by our working members. Members are getting mortgages on and off reserve. People are starting to have less dependence on council. People exposed to a good wage want nothing to do with council anymore."

He sees this as incredible progress. "Ten years ago, if a light bulb in their house was out and needed changing, they would call us because they thought that was our job."

The Haisla have a justifiable pride in accomplishment. "You come to my community and talk about potential and opportunity, and you see what we have achieved. You will see the amount of land we have acquired, the amount of revenue we have created and acquired, the services we have set up for our people and how we have restructured our governance. You will see the amount of wealth acquired by our individual members."

Chief Ross provides credit to Premier Christy Clark for assisting with his people's ability to realize their dream. "I first met Christy Clark shortly after I became Chief Councillor, in 2011, and she was the new leader of the party. It was a very positive meeting. Our council had established the idea of answering, 'How does government work?' and once we found out, we had decided we were going to work with them to get things done."

What about the LNG connection? "The first LNG proposal was brought to us by the BC Environmental Assessment office, for our input, in 2004. We thought, because we were engaging in a provincial assessment process, that BC would get on board with the proposal, but that was not the case at all, at least not until 2011.

"The proposal was brought forward by a group of people called the Galveston Group, the project was called KLNG, and really what they wanted was an environmental certificate, so that they could then sell it to the big guys, which is what they did. We didn't know that at the time. On every project that came along, we learned more about the process, and the project, and the finances, and the industry, and the liability. Next time, we will negotiate our position differently."

Ross says that they would not make the same concessions today that they made previously regarding the LNG project in their territory. "The current proponent is Chevron and Woodside, and we have a great relationship."

I ask Chief Ross why he thinks so many First Nations are angry with Premier Clark or protesting her government's initiatives, and he replies, "I ask this question to my people as much as I can. I credit industry and Christy Clark and her government for every benefit my people are enjoying now, and I tell them that at our membership meetings. I say 'this benefit is due to LNG,' and they need to understand the connection."

In talking to Ross, his sincerity and passion are clear. "When I see articles criticizing Christy Clark, when I see the stories from our people demonizing her, I say to my people that she has delivered on every one of her promises, and more."

He says she is a good premier to work with First Nations. "When I first met her she said, 'I know there are huge issues, and there has to be a better way to deal with them. I promise you, we will help you find a way for you to get out from under the *Indian Act*, to get deals done away from the treaty table and to deal with your community's needs.'

"She delivered on that, and she keeps coming through for us, time and time again."

Ross finds the premier caring and attentive to his people's needs and to their agenda. "We meet in passing at airports and conferences where she normally asks, 'How's it going? What do you need?' and my reply, more often than not, I say, 'Nothing. We are working with your staff and ministers and everything is going great.' She is good at delegating tasks to a very competent and capable team," Ross says. "She makes her promises and she keeps them."

Ross's frustration is evident in his voice when he talks about the gap between his community's experience and the public commentary. "I see First Nations condemning her and really have to wonder, what have you done to try to find a solution, a better way?

"If there is a problem between two parties," says Ross, "both parties have to work together to find a solution, to find common ground. The government can't always have First Nations issues at the top of their list, they have so many other issues to manage. It is better for us to work with government, and recognize their reality."

Is Chief Ellis Ross optimistic about the future of First Nations? "First Nations that are willing to accept that economic development can occur in an environmentally sustainable way, the ones with clear leadership like Sechelt, Westbank, Osoyoos, and others. I used to be envious of those bands because of their wealth, but you don't see them complaining, they are focused. These First Nations will be bigger, stronger, and wealthier as they move forward. I am not optimistic about those First Nations that are still complaining about the same issues they were complaining about thirty years ago."

ONE OF THE most profound comments in my interviews came from Wendy John when she explained something her grandfather told her shortly after she was elected to Musqueam council.

"When I got into politics, I really didn't like it at all. My grandfather was a big influence on me ... I remember, one day when I was driving him, he was asking me about my role, and he knew I was thinking of quitting. He had this way of talking, when he was very serious, and he wouldn't look right at you, he would bow his head and look at his hands, and when he did that this time, I had to really pay attention to what he was saying. This is what he told me:

> Wendy, you have a responsibility to help our people understand what was done to them. You also have a responsibility to help people outside understand what they have done to our people. Most important, you have a responsibility to help people outside understand what they have done to themselves."

Wendy John pauses in sharing this, as if she can still remember her grandfather beside her, and she explains that she did not understand what he was saying at the time. Fortunately, she did not quit; she persevered and later served as chief.

"This has been with me often, and I have tried to understand what he said, and I think I finally do... Prime Minister Justin Trudeau, whose comments when he took government show that he understands that a new relationship with First Nations is important for all Canadians [said] 'It is time for a renewed, nation-to-nation relationship with First Nations peoples, one that understands that the constitutionally guaranteed rights of First Nations in Canada are not an inconvenience but rather a sacred obligation.' And Trudeau repeated this in Paris, when he said that indigenous people can teach the world how to care for the planet.

"And when I heard his comments, I thought, this is what my grandfather meant: that part of my responsibility was to help people outside our community understand that they need to build this trust with us for their sake, not just our own."

WHEN PREMIER CHRISTY Clark talks about the issues that her government is addressing, she speaks most passionately about First Nations.

She sees several areas that need attention and recognizes the treaty process must be revisited. "Now that the federal government, for the first time in ten years, is interested in looking at a solution," says Clark, "we may get somewhere."

She knows that it takes all parties, First Nations, the province, and the Government of Canada, to negotiate treaties. "Treaties are the gold standard for Reconciliation."[29]

Part of her government's initiatives on Reconciliation will address returning cultural heritage items to BC, and restoring these items to their rightful owners: the Indigenous people.

The premier is very excited about this project of working with First Nations to repatriate their cultural items, which includes numerous and beautiful objects that were basically stolen from Aboriginal communities and put into museums around the world, usually without the consent or knowledge of the First Nations who owned them. This includes everything from carvings to baskets to clothing items.

When Premier Clark talked about this project, she banged the table with her passion to get the items back to their rightful owners. Her focus is sincere and she is on a mission. She said, "There are two things that have driven me on this. Initially, in talking to my First Nations friends and colleagues, I've really grasped how important, spiritually and culturally, these items are, and how many of them have been stolen. It is an *essential* part of Reconciliation that these items be returned. But I'm also driven because these items belong to First Nations first, and they belong to British Columbians second, because they are part of our shared history."

She believes the restoration of Aboriginal cultural items is important for all British Columbians, "The Europeans treated First Nations culture as if it was a dying culture needing to be preserved under glass; they spirited these objects away to museums around the world. First Nations are not some kind of Neolithic civilization that left a few arrowheads behind! They still exist, these cultural items are still relevant, and *they want them back.*

"And I want to help them get them back."

Listening to her, I have no doubt she will succeed. If you are reading this and you have some of these items, I suggest you send them back right away.

As for the legal standing, Premier Clark recognizes that the Tsilhqot'in Decision enhances Indigenous rights and title, and sees this as additional clarity in working on land claims.

Clark has been criticized for her heavy focus on resource projects, including Site C and mining, and most particularly her emphasis on the development of an LNG industry in BC. She sees First Nations economic independence as intrinsically linked to their ability to break away from historic injustice and the *Indian Act*. She remembers meeting Chief Ellis Ross of the Haisla, and his passionate pitch for her support of LNG for his people.

Says Clark, "When I met him for the first time, he started talking to me about why LNG was so important to his community. He said it was critical that they create their *own* economy and their *own* future based on their *own* experience and their *own* vision."

Premier Clark does not believe that it is all about money for First Nations; it is something deeper, and she thinks sometimes industry partners need to understand that. "I think the relative value of money is often not understood when working with First Nations. You hear a story about how a company went in and offered them a whole bunch of money. Why didn't they take it? The answer is because they see resource development and the wealth that it creates through a different lens."

She supports their right to view the projects through their historic lens. "In order to make good decisions around resource development, we have to work with them to make sure we develop resources in the most responsible way."

Why is she moving ahead with LNG when so many environmentalists and First Nations are speaking out against it? "One of the things that is, to me, so important about LNG is that it will give us the opportunity for the first time in over a hundred years to allow the First Nations to get in on the ground floor on the development of a resource at a time where the pie is uncut."

Premier Clark is hearing about the need for economic development from many people in the First Nations community, and offsets the negative commentary against this because she wants First Nations leaders to be at the table throughout the industry's development and make sure they do not arrive later, when they would have to fight for a share of the wealth generated.

"That is always the problem with trying to allocate part of the pie later. For all the other resource industries, First Nations are trying to negotiate a piece of a pie that's been divided for decades. LNG is a once-in-a-generation opportunity to change things forever."

When Premier Clark speaks on the opportunity for remote Aboriginal communities, she becomes quite excited about how transformative the opportunity is, not just in terms of the wealth generated, but also in terms of Aboriginal people being able to safeguard their environment and ensure sustainable decision making. Also, moving into partnerships with government and industry provides a chance to leave behind the generational social problems and become successful decision makers.

"I feel so passionately that we have to take this opportunity, when we think of our legacy as British Columbians, I think that this generation should feel so proud to say, 'I was part of the generation that changed the trajectory of poverty and suicide for First Nations in BC.'

"First Nations communities are really ready. After a generation of so many of them being able to go out and get the education they need to be the leaders their communities need, now they can step up to take advantage of this opportunity, and look forward, and set up a legacy.

"Their predecessors didn't have that chance."

Yet there is a light I love, and a food, and a kind of embrace when I love my God ... where there is a perfume which no breeze disperses, where there is a taste for food no amount of eating can lessen, where there is a bond of union that no satiety can part. That's what I love when I love my God.

ST. AUGUSTINE

CHAPTER 24

OPPORTUNITIES

ASK PREMIER CLARK, "As you fly over BC and look out the window, what do you see?"

She replies, "I see the astonishing beauty of the natural world and the diversity of it in British Columbia, and I also see the ingenuity and resilience of the people who live here and who made this place. When I drive through a smaller community, where things can be more visible, I wonder how people carved life out of these remote areas. BCers are really resilient and ingenious. There's no other way to carve out a life in a place like this."

As she describes this, her respect for the efforts of these residents is obvious. The Interior of the province, where much of the resource wealth originates, is pretty rugged, with vast tracts of wilderness.

"We have a lot to protect and preserve," says Clark. "We are so lucky to live here. I never forget how lucky I am, we all are, to be able to live here.

"As premier, I see it as a place of huge opportunity, and my job is to help as many people as possible be able to make something of those opportunities. I need to help them make the most of the opportunities that are so abundant in this province.

"Thinking of a single mom in Surrey, what does she need that she doesn't have in her life? What can I do that can make it possible for her

288

to take advantage of what we can offer? Education, absolutely. How do I help her obtain job security? How do I support her in raising the best kids that she can?

"This is what I think about as premier: people are resilient, people are ingenious, there's lots of opportunity, but some people need help realizing the opportunity, and that's where I come in."[1]

One of the premier's first initiatives when she became leader was the BC Jobs Plan, introduced at the Vancouver Board of Trade on September 22, 2011.[2] This was a three-pronged plan comprising job creation, which includes government facilitating business development through tax and regulatory changes; building smart infrastructure, including investing in people training and job-creating infrastructure projects; and opening up new markets for BC goods, especially in China and India, where the middle class is growing rapidly.

In the first few years of Premier Clark's government, there were many news stories saying that her jobs plan was failing. In January of 2014, a high-profile editorial by Iglika Ivanova was published in the *Vancouver Sun*. It was based on Ivanova's report entitled *BC Jobs Plan Reality Check* written for the Canadian Centre for Policy Alternatives. In her report, she examined some labour market performance indicators since the September 2011 Jobs Plan announcement and compared them to the economic performance in the two years that preceded the Jobs Plan.

The findings reveal a largely jobless recovery, which was not significantly boosted by the BC Jobs Plan. While the jobs recovery has been disappointing across Canada, B.C.'s is weaker than the Canadian average. The persistently high unemployment rate, still over two percentage points above the pre-recession levels, is only the tip of the iceberg.

The Jobs Plan was supposed to stimulate private sector job creation, but the private sector actually lost 12,000 jobs in the first 10 months of 2013. It's very rare for the private sector to shed jobs outside of a recession. In the last 40 years in B.C., it has happened only once, in 2001, and then only about 2,700 jobs were lost, much fewer than last year.[3]

There were a number of similar negative stories published in the years following the announcement; however, in January 2016, Victoria *Times Colonist* columnist and press gallery veteran Les Leyne wrote a column with a different take on matters. It was entitled "B.C. Job Stats Not Great, but Not Bad."[4] In this column, Leyne refers to a Statistics Canada report on BC's position in Canada in jobs in 2015.

> B.C. had the fastest employment growth in the country. The employment rate rose 2.3 percent (52,000 jobs). Employment has been on an upward trend since April. The 52,000 jobs is almost a third of the 158,000 new jobs created nationally last year.

Although it is arguable that governments cannot always take credit for job rates, given that the global economy often has a direct impact on job creation and job losses, government can create public policy that creates the business environment for jobs, or one that obstructs investment and, therefore, creates an environment where some jobs are scarce.

In 2016, there are a number of very large infrastructure projects in BC poised to create jobs, at the same time that BC has its fourth balanced budget. Whether these jobs go to British Columbians depends on factors including whether or not the labour market can provide workers with the necessary skills and whether or not there is a plan in place to encourage the proponents, or employers, to hire BC companies or workers. All of these factors have many elements that require discussion to understand whether or not Christy Clark's government can take credit for BC's job performance in Canada.

One large infrastructure project that Premier Christy Clark announced on December 14, 2014, was the Site C dam project in Northern BC's Peace River area. The Site C hydroelectric project was first proposed in the mid-fifties as part of W.A.C. Bennett's energy infrastructure, the third of four proposed dam projects, and has been discussed by successive governments without a decision. In 2010, Premier Gordon Campbell agreed to initiate a regulatory review,[5] the results of which were inherited by Premier Christy Clark, who agreed to review the proposal in detail after her election in 2013.

"It was a really hard decision from a lot of perspectives," Clark says. "We spent a year looking at the finances and all the details. There were First Nations issues, there were agricultural issues, workforce issues, Hydro ratepayers. We had to make sure we didn't rob the emerging LNG sector of the well-trained workers. We had to make sure we didn't inflate the cost of labour so much that it was impossible to do business here."[6]

The cost of the project is estimated to be approximately $9 billion, although that number may increase as the project's start is delayed. BC Hydro is pushing for the project in anticipation of future energy demands that BC Hydro does not believe can be met as effectively by other technologies. It is a highly controversial project, uniting environmentalists, agricultural activists, and some Aboriginal people in opposing it.

According to a CBC story,

> B.C. Hydro forecasts it will need additional sources of electricity by 2028. It purposely calls the dam, the "Site C Clean Energy Project" claiming it will produce fewer greenhouse gas emissions for the amount of energy supplied than any other source except nuclear power.
>
> The dam reservoir would require the flooding of approximately 5,500 hectares of land and more than 83 kilometres of river valley along the Peace River and its tributaries... This would include over 3,000 hectares of wildlife habitats, heritage sites, and Class 1 and Class 2 agricultural land.[7]

At the same time that Site C united many people in their opposition to the project, it caused divisions in the Official Opposition NDP caucus, with environmentalists facing off against pro–resource jobs MLAs. Billed as Premier Clark's attempt to create a "wedge issue" in the NDP, Site C is likely to be part of the big debate in the 2017 election. In politics, a wedge issue is one that is used to divide a political party and is seen as political gold to the party that is united. If Clark succeeded in creating a wedge issue in the NDP with her announcement on Site C, that was likely a bonus for the Liberals. The *Globe and Mail*'s press gallery reporter Justine Hunter captured the issue well in an October 2015 story.

When the B.C. New Democratic Party caucus met over the summer, it faced a divisive resource debate that culminated with at least one MLA threatening to quit.

The topic was the $8.8-billion Site C dam. The New Democrats under the leadership of John Horgan want to rebrand themselves as a job creation party. By supporting Site C, the largest public infrastructure project in the province's history, the NDP would have had a chance to win back the support of the building trade unions that abandoned them in the 2013 provincial election.

Mr. Horgan, as he prepared to take over the leadership of the party in the spring of 2014, was aware of the possibility of being boxed in, again, by the B.C. Liberals, who successfully framed the NDP over its opposition to the Kinder Morgan Trans Mountain pipeline proposal during the last campaign.

But on Site C, the environmental wing of the caucus would have none of it. Lana Popham, the NDP's agriculture critic, spent part of her summer paddling the Peace River in solidarity with opponents of the project. When it is complete, the dam will flood more than 4,000 hectares of once-protected farmland. Sources say she made it clear she would quit the caucus rather than vote in favour of the project. She wasn't alone. Other influential voices including George Heyman and David Eby stood with her in arguing against the project. The critics of Site C won the debate but the NDP did not take a public stand on the project until last week, when the B.C. Liberal government forced it off the fence by tabling a resolution in the legislature that required MLAs to vote yea or nay.

The resolution was designed to allow the Liberals to play wedge politics. Premier Clark and her caucus had buttons printed and ready to wear, declaring their support for the project. Once the vote was tallied, the premier was moving her ministers out of the way in Question Period to deliver her sound bite of the week, about "this Opposition's opposition to jobs."

Tom Sigurdson, head of the B.C. Building Trades, got a heads-up call from the NDP before the MLAs filed into the House to vote. It was a courtesy call but he did not hide his disappointment, making it clear his members will remember this when the next election comes around. "I said, 'You've got to come up with a jobs plan.'"

Then again, the Liberal government has failed to deliver so far on jobs for his members, insisting on an open shop labour model for the project that so far has resulted in almost half of the jobs going to workers from outside B.C. "There are a lot of Alberta plates parked at the construction site, and my members are unemployed and looking over the wrong side of the gate."

So the wedge is not firmly secured yet, and Mr. Horgan maintains he still has plenty of time to develop a jobs plan that will bring the party's traditional labour allies back into the fold before the May, 2017 election.[8]

The first two hydroelectric projects in northern BC transformed their regions, including displacing the Sekani First Nation. In 1992, when I was an MLA, I visited Williston Lake, which was created by the W.A.C. Bennett Dam, to learn about the area, courtesy of forestry giant Fletcher Challenge.

While I was on this trip, we visited the new village for the Aboriginal people. While my colleagues were occupied somewhere else, I walked into the residential area, where two Sekani elders were sitting on the steps of their new home in the sunshine. I sat with them for a little while, and they quietly shared some of their stories of life before, during, and after the flooding of their village and the loss of their traditional life. It was a moment of profound sharing, and I can still remember feeling the loss of their lifestyle and lifelong home as they told their sad stories through gentle remembering. A couple of small children rode bikes nearby in the lane as I listened and they talked. Afterward, I toured the beautiful new school that had been built by the forest company with the support of the government, and as I toured it, I was still seeing the lost village.

On a map of BC, Williston Lake looks like a long blue line, and you cannot appreciate its immensity without visiting it. I remember standing on one arm of the lake, and it was so big it was like standing on the edge of an ocean. Our guide told us that there were still trees, called snags, at the bottom of the lake, and the boats all needed steel bottoms because if a tree comes loose, it shoots to the surface and can puncture the hull of a boat. There is a vastness to the sky and the water in this part of the world; the air is fresh, the forests stretch for miles, and the world feels new.

Regardless of the history of hydroelectric projects in BC, and the social and environmental impacts that are beyond the scope of this book, it is indisputable that residents and businesses of our province enjoy clean and affordable energy sources because of the BC Hydro electrical projects initiated by the W.A.C. Bennett government.

Site C had to pass the environmental impact assessment process, and it did, prior to the government's approval. When I interviewed the premier about Site C, I asked her how it was possible for something that is so clearly damaging to the environment to be cleared by an environmental impact assessment.

She replied, "Every development has a cost/benefit analysis, and every action has an environmental impact. With Site C, the benefit won. This project will ensure an abundance of the cleanest power in the world for another generation of British Columbians."

Clark points out that though there is an environmental cost to pay for a project of this magnitude, "there is no cleaner, more affordable source of power. You shouldn't be growing an economy based on coal power."

What about technology, and the development of alternative options? "Site C grows the tech industry, enables us to have electric cars, it supports all of our industry. We are going to need the power," says Clark, "so where do we get it?"

Premier Clark has been vocal about the need for choices that have less impact on climate change, and her government's priorities on energy: "We choose clean. Our electricity in the province is already 97.9 percent renewables. Two-thirds is hydro, one-third is independent power."

Making a decision to proceed with Site C fulfills Clark's first and second commitment in her jobs plan because BC Hydro had been arguing for approval of Site C since the early 1980s, with government refusing to make a decision. Making a decision fulfills her commitment to "get out of the way," which is the first part, and infrastructure investment is the second part. However, in order to have British Columbians with the skills to fill the jobs that will be available through Site C, there has to be a strong commitment by government to skills training.

Has there been a concerted effort to direct money toward skills training? Yes, absolutely, and it has had its share of controversy. In April 2014, Premier Clark's government announced BC's Skills for Jobs Blueprint: Re-engineering Education and Training, a transformative program dealing with educational offerings for young people in BC, from a young age through post-secondary education.[9] Three cabinet ministers, Shirley Bond, Minister of Jobs, Tourism, and Skills Training; Peter Fassbender, Minister of Education; and Amrik Virk, Minister of Advanced Education, all made the announcements, speaking to their various responsibilities. John Rustad, Minister of Aboriginal Affairs, also attended because of the strong focus on providing access to the new program for Aboriginal students.

Minister Fassbender said, "Poets are still welcome in British Columbia's plan to re-tool the education system from kindergarten to post-secondary institutions, but more welders would be nice."[10]

The controversy occurred because the funds being directed to this re-engineering were coming out of the existing advanced education budget. The *Globe and Mail* story included a reference to efforts made by the universities to prevent the diversion of funds.

Universities argue that the pursuit of knowledge and critical thinking is a valid education goal. The province's six research universities made a joint presentation last fall to the government's finance committee, in which they said the province needs "innovative thinkers with the ability to transform and adapt to new technologies [and] a fast-changing marketplace and global environment." The university presidents stressed that employable skills are not always as clearly defined as a welder's ticket.[11]

The business community had identified the need for training as soon as the premier began talking about introducing the LNG industry in BC. According to Jock Finlayson, executive vice president and chief policy officer of the BC Business Council, the need for training was an issue across Canada. In a blog story published in April of 2014, just prior to the government's announcement, Finlayson shares data from a number of major business organizations including the Canadian Chamber of Commerce, the Canadian Federation of Independent Business, and the Canadian Council of Chief Executives (CCCE), all identifying skills shortages as an issue for their members.

While academic researchers and policy analysts continue to debate the extent and implications of skill shortages, employers in Canada seem convinced that shortages exist and are an important factor constraining business expansion.[12]

British Columbia has a long history of wealth generation through resource-sector investment, whether forestry, mining, agriculture, fisheries, energy, or other resources. In recent decades, the majority of voters live in urban areas and are not connected to the resource industries that generate BC's wealth. This urban vote concentration meant that many political leaders moved away from policy that dealt with the needs of these hard-hat industries.

Employers in many of these sectors claimed there was a shortage of skilled workers in areas like welding, and a new LNG industry only adds to the demand. Trade organizations for the workers were also very concerned, as Vaughn Palmer captured in an April 2014 article entitled "On LNG Skills Training, Labour Works the Apprenticeship Angle," wherein he talks about BC Federation of Labour president Jim Sinclair's reintroduction to the halls of the legislature after years of Campbell's government. The Clark government invited Sinclair to be part of the working group that put together the fifteen-point final report on training and recruitment needs for BC's workforce, and he attended the release of this report along with representatives of government, industry, and First Nations.[13]

According to Tom Sigurdson, executive director of the British Columbia and Yukon Territory Building and Construction Trades Council (BC Building Trades), "The skills training initiative was because there was a confluence of events, issues, and organizations that had the government recognize that those people who were investing had to have a skilled work force before they would commit to large investments. You look at LNG Canada for example, of which Shell is the major shareholder, before they invest billions in an energy project, they need to trust the skills of the workforce. They are looking for masters of trade, not jacks."

In any conversation with Tom Sigurdson, it is obvious he knows his topic. He understands the dynamic between politics, industry, and the workforce, and is passionate about the needs of the workers. Many labour leaders will put politics before their members; Sigurdson consistently puts his members and their interests first.

"The construction trades need a steady supply of trained workers. If anyone wants to take on an apprenticeship, they need access to sixteen hundred hours of tool time, which is forty weeks of work after six weeks of theory. If they can't get this, they might end up working in the underground economy, or they find work in other industries and leave their apprenticeship quite frustrated."

Sigurdson's frustration is obvious in his voice as he explains that the need to provide the hours of tool time is just as important to the industry as it is to the worker. Although the climate between private sector unions and the government improved under Premier Clark, Sigurdson is careful to point out how, in his organization's view, there are still significant problems that affect the industry, and the workers.

When Tom Sigurdson was asked what kind of a premier Christy Clark is for the workers, he provides an insight into the private sector union's perspective on the state of worker protection. "I have to say there have been no amendments to the Labour Code, and no consultation asking us if we need amendments to the Labour Code. They are not taking a look to try to rectify the damage that was done by gutting the Labour Code under the Campbell administration. It is very difficult to work with. We have workers who sign union cards, and we try to represent them, and by the time we get in front of the Labour Board, we lose the vote because of the employer intervention."

To be fair, Clark did not commit to overhaul the Labour Code, and before the 2013 election, her troops were not interested in it, and the unions who were interested all supported the NDP. Still, Sigurdson believes that government could investigate possible changes. He outlines the specifics of what the union faces: "You can end up with 80 percent of a work force signed up, and then they face intimidation and fear of losing their jobs, and they can't afford to risk that, so by the time we hold the vote, the vote is lost and the attempt to certify an employer is defeated."

Is the union advancing certification out of self-interest? Sigurdson says, "In 90 percent of the cases, the workers are the ones pushing to start a union. We are not out there trying to sell the idea to them, they come to us because they feel they have bad working conditions, and they need help."

"Prior to the election of the Campbell administration, certain trades were designated compulsory—electrician, plumber, gas fitter, et cetera. You had to be either certified or an apprentice to do work related to the trade if the trade was compulsory."

For the BC Building Trades, it is not just about the workers' opportunities to succeed, it is also about a wider topic: safety. When talking about the need for tool time and experience, Sigurdson points out, "Much of this was about public safety as much as it was about passing on the skill-sets to apprentices. Campbell gutted the compulsory trades in BC, and it has caused problems for the trade and indeed compromises public safety."

The ability to ensure that jobs go to BC workers can be problematic, even when a premier makes a high-profile commitment, and the fallout can be loud. In December of 2015, over the Christmas holidays, the BC Federation of Labour's (BC Fed) new president Irene Lanzinger criticized the government for ignoring BC workers, joining the NDP's opposition to the announcement from BC Hydro that the Site C dam jobs could not be guaranteed to BC workers.

Reporter Wayne Moore's story in *Castanet*, an online news site in the premier's riding, quotes Lanzinger: "[T]he BC Fed says Premier Christy Clark has abandoned a 'B.C. First' policy on jobs in awarding a $1.75-billion project for the controversial Site C hydroelectric dam."

According to Vaughn Palmer, Lanzinger was the labour vice-president of the NDP in the spring of 2014, just before she became leader of the BC Fed. She also previously served as president and vice-president of the BC Teachers' Federation, coincidentally during the time period when Christy Clark was in cabinet in the Campbell government.

Moore goes on to report on Lanzinger's concern:

> The project, the single largest construction project in B.C. history, has been touted to create as many as 1,500 jobs at peak construction, with no guarantee that the jobs created will go to British Columbians.
>
> "Christy Clark has chosen an ironic moment, just before the holidays, to let B.C. workers know they won't be in line for jobs in this huge project."
>
> "Her holiday message is corporations come before the needs of workers."[14]

Lanzinger's release against Clark was in response to the award of the contract announced by BC Hydro's spokesperson David Conway, who stated that there could be no guarantee that these jobs would go to BC workers, even though he did say all efforts would be made toward this goal.

These challenges of placing workers run across the resource sectors, although first the jobs must be there in order for there to be an opportunity to fight over them. According to Jas Johal, of the BC LNG Alliance: "It is estimated that each LNG project will generate 4,500 to 7,000 workers, which is in addition to the jobs for construction of the pipeline plus all the indirect jobs such as food supply, housing, etc."

These are staggering numbers, and in fact, jobs in LNG were created within six months of the election. In September 2013, Les Leyne wrote an article for the Victoria *Times Colonist* quoting the premier as she expounds on the benefits of the LNG initiative.

> Many view the LNG concept as a far-off vision, but Clark stressed that some of it is happening here and now.
>
> "I've got news for the pessimists," Clark said, listing the current situation. One company has invested $800 million on an LNG plant site, with 500 people working on it.
>
> About $7 billion has already been spent on securing rights and opportunities, she said. Seven projects have applied for export licences, three of them approved. Port Edward, a village just south of Prince Rupert, has sold municipal land for development for the first time in 12 years.[15]

Johal notes the visual connection between the premier's commitment and what people see on the news. "One thing you will notice is that at many announcements, organized labour is present. Christy Clark is turning the Liberal Party into the hard hat party of BC."

There was an additional level of challenge to ensuring jobs for British Columbians in a new liquefied natural gas industry: the ownership of the process by global energy giants and their in-house skilled workers.

The global liquefied natural gas industry, similar to global oil or large mining, is known for its so-called fly in, fly out (or FIFO) workers. These highly trained workers move as part of the investment package, sent from country to country depending on the project location. The danger to Premier Clark's vision of an LNG industry in BC, in terms of jobs creation, is that if a proponent like Petronas or Shell comes into BC with billions in investment, they may arrive with hundreds of their own FIFO workers. The project still generates millions of dollars in revenue to the provincial payroll, but it may not employ BC workers.

How can a provincial government tell these multinational giants to hire BC workers, when they may be set up with their own trained workers, and when BC workers may be more expensive or require training?

Other challenges include housing and feeding an enormous number of workers in remote locations, ensuring there is adequate transportation in place for planes, trains, and automobiles to service these massive projects, and trying to address issues like a spike in local costs of living such as housing and food, as competition increases.

How does government try to ensure BC businesses benefit from these purchases and investments? These problems took over the premier's thinking when she was trying to decide how to deliver on her jobs promise.

"My brain is always engaged, at least part of it, in trying to work through the latest problem. In a big project, for example, the liquefied natural gas, there is always a problem that needs to be worked through at the negotiating table. There are always issues that need to be resolved. When I get a chance to sit quietly, those things will come to the fore, and sometimes an answer will present itself quite quickly because I've been thinking about it in the back of my mind. Issues like First Nations issues need focus, you have to turn your mind to it and really think."

Tom Sigurdson's 35,000 unionized members in BC's construction industry are keen for jobs. He is justifiably cynical about political messages about job creation. Sigurdson met Clark briefly when she was in the Opposition, but it was just an introduction to building trades. When he met her in the fall of 2013, it was significant because she had opened the door to organized labour, welcoming them back to discussions with government after a dozen years of being excluded from any decisions.

"At that time," says Sigurdson, "the premier and the government were still basking in the 'glow of LNG.' Many of us in construction were not as optimistic about how big the LNG opportunity was or is going to be. We have had lots of proposed projects never see the light of day.

"During the election, there had been talk of a Prosperity Fund, and tens of thousands of jobs in BC. It was really over the top. I never did subscribe to the enthusiastic side of the LNG conversation as presented by the premier. I knew there would not be eighteen proponents shipping LNG from here."

In the early speeches the premier made about LNG, and throughout the election, she talked about a number of large projects, and the size of the opportunity if all BC proposals happened. Sigurdson says he didn't buy in, even at that time. "Still, it didn't have to be as big as it was presented to be good for the workers and good for BC."

Following the 2013 election, global oil prices steadily declined, and in the race to own the emerging LNG sector, the United States and other countries outpaced BC's rate of approval. In BC, environmental regulations and First Nations consultation slow the process down.

Says Sigurdson, "Expectations have certainly been tempered since the statements being made in the election. LNG is not the great promise it was in 2013. But it doesn't have to be. If one project goes ahead like what is being discussed in Kitimat or Prince Rupert, we will see twelve thousand jobs in the construction alone, and fifteen thousand workers dedicated to the success of the project."[16]

The expectation that projects in BC will mean jobs for BC workers is a critical political point. The problem of huge investment without jobs is not one that is easy to solve.

"In order to solve any problem or come up with a solution," says Clark, "I have to put myself in the other person's shoes. I have to really understand who they are, how they came to express the opinions they have, and when I do put myself in that perspective, then it almost always helps me come up with what we can do."

It is important to try to understand all the angles. Says Clark, "Someone in the business world told me that there's no such thing as a deal with only one winner, because that's a bad deal, and it won't last. The only way to have a good deal is to have both parties come away as winners. You can't divide the world into winners and losers.

"I think that's part of the problem with those who are really ideological about politics, because it's all about 'I'm going to win and you are going to lose.' The politics of division that we see with people who are ideological is a mug's game."

She decided that a persuasive approach to hiring BC workers had to be part of the strategy in developing a new energy industry, and a dialogue on this had to occur while recruiting international investors. "You can only build a good, healthy society if people feel included. People need to feel part of the win."[17]

The Union of British Columbia Municipalities (UBCM) is the largest convention of local governments in BC every year. It is often the site of major announcements by the provincial government. In the fall of 2013, about four months after the general election, Les Leyne reported Clark's announcement of the LNG Buy BC Program to connect multinational corporations investing in LNG with BC workers and suppliers, with a comment from the premier about the intent of the program: "We are going to ensure that as many benefits as possible flow from the resource to the owners of that resource ... the people in every corner of this province."[18]

Robert Barron of the *Nanaimo Daily News* reported more details on the program as the premier unveiled them.

Clark also said the government's new "LNG Buy B.C." program, announced last month, will play a major role in helping B.C. seize the opportunities presented through natural gas. Under the program, the government will act as a "matchmaker," connecting B.C. businesses to the multi-national corporations behind the proposed LNG projects in the province. Clark said British Columbians have an "incredible chance" ahead of them to create thousands of jobs, pay off the province's debts and become the biggest economy in Canada.[19]

The new program would "matchmake workers with opportunities"; however, the premier spoke of the need for relevant training.

> Clark said the forecast is for up to one million new jobs to be created in the province by 2021 if the economic opportunities related to natural gas exports are realized, including 22,000 workers that will be needed to build the pipelines alone. But she noted that approximately 20 percent of the students in the province's education system don't graduate, and that number is much higher for aboriginal students, so much more work and effort is needed to keep kids in school and train them for future opportunities.[20]

On October 30, 2013, Clark made the announcement of her choice for the position of LNG Buy BC Advocate. The appointment surprised observers and came with a wide range of responses. I will deal with *that* in the next chapter.

Deputy Premier Rich Coleman is the Minister of Natural Gas Development and Minister Responsible for Housing, and is tasked with establishing the LNG industry in BC:

> This is going to be one of the most dramatic shifts in economic development in my lifetime. The First Nations in northern BC don't have economic opportunities, they have a bit of forestry and a bit of mining. LNG is a once in a lifetime opportunity for them to change their economic reality. I have heard some of the First Nations Chiefs stand up and talk about the need for opportunities for the young people in their communities. It's quite emotional. The training, jobs, and social outcomes in these areas will be significant, and their leaders see this.

The premier and Coleman face some pressure from proponents of the initiative. Jas Johal of the BC LNG Alliance works with all the key industry players and has a close perspective on activities in this emerging sector.

"The premier made the LNG industry a key part of her campaign focus," says Johal. "After the election, there were a number of people who tried to say that she hadn't delivered, but they have no concept of the reality of the industry. In fact, the LNG industry proposals have moved quickly in the first two years. All projects are moving forward at their own pace, and these are massive projects. It is true that global energy prices, and their dive in 2015, created a 'headwind' to progress, but these industries have ups and downs and that is the nature of the business."

Liberal Party president Sharon White was clear on the government's intentions. "I think LNG is an opportunity for transformational change for the province. I think it is real, and it will happen, and I think it shows the premier's ability to seize on opportunities. It will give people in smaller northern communities hope that they can have good long-term jobs."

Is LNG a good opportunity at any cost? White admits there must be safeguards. "I think that in BC we should expect and demand that we meet higher standards than anywhere else in the world. We will carry out our resource extractions in the most responsible manner possible. Our citizens expect and demand this of us."

That wise man of BC politics John Reynolds put it pretty clearly. "If you have a good economy, even a union guy is going to vote for you. They want jobs. Look at LNG, that's jobs. The NDP are opposed to a lot of these ideas, and so people don't want to vote for them."

The NDP have taken a position generally in favour of LNG, even as they oppose the Liberal government's tax structure and some of the specifics of deals under way in 2016. Vaughn Palmer gives us some insight in a June 2015 *Vancouver Sun* story:

> With the B.C. Liberals preparing to recall the legislature to approve a project development agreement for liquefied natural gas, Opposition leader John Horgan is signalling that the New Democrats will likely be voting no.

"We'll have to take a good hard look at what they've got on the table," he said, referring to the government promise to release the full text of the agreement with the Petronas-led consortium that is proposing a $36-billion LNG development in B.C.

"But based on what I know, I'm pretty disappointed with where we've gone on this file," Horgan told me during an interview last week on Voice of B.C. on Shaw TV. "There are a whole host of giveaways from a government desperate to get a deal at any cost."

"Any tax changes that have an impact on Petronas will be frozen from this point until the end of the project," complained Horgan. "I didn't sign on for that. I signed on for getting the most value we could for the people of B.C. and what the Liberals have done, is focused on the needs of Petronas."

Horgan "signed on" to LNG last fall, when he led the NDP caucus to vote in support of the government's proposal for a new tax on the production of natural gas in liquefied form.

By his own admission, Horgan got "a lot of grief from within the NDP and outside the NDP" for that bipartisan gesture. It put him in the position of supporting fossil fuel development in general and natural gas fracking in particular.

But having insisted on assuming the NDP leadership in 2014 that the party most needed to develop a credible platform for economic growth, Horgan has continued to defend LNG development as long as it delivers a fair return.

If that expectation is borne out, he'd risk undermining his other stated reason for supporting the LNG tax—to refute the Liberal propensity for saying the NDP is against all development. "You won't be able to say that with this bill," Horgan told the Liberals at the time of the vote, "because we're going to stand side by side with you and vote in favour of it . . . As deficient as it may be."[21]

Jas Johal offers a different perspective. "Given that the bulk of BC works for the private sector, why is the political messaging so negative against corporations? For example, when the premier introduced the

LNG tax, there were howls from the NDP that the tax was too low; however, this tax is on top of the royalty payments, BC's carbon tax, plus corporate taxes, municipal taxes, payroll taxes, GST, and every other tax. BC is the only jurisdiction in the world that we know of that even has a tax for LNG. Still, the proponents are here."

The publication *Alberta Oil* did a feature piece on Premier Clark entitled "Like Her or Loathe Her, B.C. Premier Christy Clark Is Possibly Canada's Most Influential Energy Player." Written by Canada West Foundation senior fellow Dr. Roger Gibbins, the essay is such an effective outline of the issues that I have provided significant excerpts here:

> British Columbia Premier Christy Clark embodies the tension Canadians feel between energy development and environmental protection. The fate of her political career depends on finding the right balance...
>
> The centerpiece of (Clark's) vision is the prospect of liquefied natural gas exports to Asia. Although B.C. gas supplies are huge—perhaps enough to meet Canadian requirements for 150 years, according to recent National Energy Board estimates—North American markets are saturated and prices are chronically low. However, Asian markets are expanding, and the large price differential between Asian and North American natural gas markets is enticing. Asian markets are relatively close, and LNG presents lower environmental risks than does the tanker transportation of oil.
>
> It can also be argued that LNG exports would reduce global GHG [Greenhouse Gas] emissions if they displaced Asian coal consumption...
>
> However, realizing the LNG vision will not be easy...
>
> Nor is time on the premier's side, for the LNG world is moving very quickly. The competition for long-term Asian contracts and investment is intense, and B.C. producers are the new kids on a very tough block...
>
> Premier Clark's political challenges are more difficult than those faced by Alberta premier Alison Redford. Clark confronts a more fractious and ideological electorate, beginning with a deep division

between the urban south coast and the resource-dependent interior. Alberta's urbanites are generally on-side with resource development whereas the metro Vancouver electorate, many of whom believe civilization stops at Hope, is disengaged at best and antagonistic at worst.

no Alberta mayor offers the implacable resistance posed by Vancouver Mayor Gregor Robertson.

The B.C. environmental community is not only larger and stronger than its Alberta counterpart; it is also more radical, often promoting zero-growth, the aggressive pursuit of a low-carbon economy, and a strong commitment to domestic initiatives to mitigate global climate change. In addition, the First Nation landscape in B.C. is much more complex given both the lack of treaties and vigorous, entrepreneurial First Nation leadership.

When Premier Clark identified significant economic benefits as one of her conditions for supporting the proposed Northern Gateway pipeline, and presumably the twinning of Kinder Morgan's Trans Mountain line, she neatly captured a widespread perception that the economic benefits of oil pipelines would flow almost entirely to Alberta...

As a consequence, there must be times when Clark wishes that Alberta and its bitumen would just go away... If Clark could throw oil and oil pipelines under the bus, she would do so...

Thus the delicate balancing act Clark faces is not to go offside with her own voters while at the same time demonstrating to potential investors that B.C. is open for business...

Premier Clark's demand for a fair deal for British Columbians with respect to the proposed Northern Gateway pipeline hit a very responsive note in the province...

However, the premier's demand could violate the principled foundation of the Canadian economic union by creating new internal barriers to trade... Again, then, Premier Clark finds herself on the high wire, balancing B.C.'s position as Canada's gateway to Asian markets and investment with the isolationist impulse that is never far from the surface in B.C. politics.

The B.C. environmental community has numerical and financial strength, close ties to well-heeled celebrity environmentalists in California, and has signaled that it will oppose virtually all forms of resource...

At present, environmentalists and First Nations may appear in lockstep when it comes to opposition to oil pipelines, but First Nations face a more complex set of tradeoffs than do environmentalists.

While the lifestyles of metro-Vancouverites would take a real hit if the resource sector tanks, this possibility is either ignored or discounted in the environmental community, or at the extreme is even seen as a good thing in the context of global environmental and social justice objectives. At least potentially, however, First Nations could be major beneficiaries of resource development as gas deposits, pipelines and terminals all touch upon First Nation land. So far, Clark's LNG vision has been met with cautious First Nation support rather than opposition...

The issue is again one of balance, and Christy Clark will play a critical role in finding a First Nation balance between environmental values and economic development within her LNG vision. If this balance can be found, the political alliance between environmentalists and First Nations would be weakened.

It won't be easy. Premier Clark faces a very complex set of energy challenges. Without question they will tax her very considerable political skills to the limit. If she falls from the high wire, the consequences will be felt well beyond British Columbia. Her success or failure will shape Canada's energy sector, and destiny, for decades to come.[22]

Jas Johal sees the premier's ambition as a monumental undertaking and believes she has a workable plan.

"If Premier Christy Clark can develop a brand-new resource industry in BC, she will have left her mark on not only BC but also on Canada. She is coordinating this development at all three levels of government plus First Nations, industry stakeholders, unions, educational programs, training, plus the building of pipelines through two sets of mountain ranges."

To put it in context, Jas Johal says, if Clark succeeds in building the LNG sector, "she will have a legacy like that of W.A.C. Bennett: think big, build big."

Love is surrender. You expose your deepest vulnerabilities
and give up your illusions of self-mastery.

DAVID BROOKS

CHAPTER 25

THAT WILSON, AGAIN

C HRISTY CLARK: *Behind the Smile* is my third book about politics in
British Columbia, and in writing it, I have faced the same issue
that came up in my previous two books: I am in the book and have
some interest in the subject matter, so I have to be careful to present
information as objectively as possible and to be as fair as possible to all
the people and events while still providing an intimate and human perspective that allows the reader to feel like she or he shares a close view of
some of the topics.

I try to make politics interesting and fun.

However, in this book there is another layer of difficulty for me as an
author, and that is with respect to my husband, Gordon Wilson, who is in
this book both as an independent political player and as my husband. I
have not allowed Gordon to read the book in advance or know too much
about it; however, we did get into a bit of an argument about how I would
handle him in the book. He was insisting that I take out large sections so
that he would not appear in the book too often, worrying it would dis-
tract from the story of Premier Christy Clark. However, I cannot delete
him from the book and tell the full story.

The truth is, Gordon Wilson is threaded throughout Premier Clark's
political life, from 1987 on, from his knowledge of her father through

Clark's early days in the party, their antagonism when she was in Opposition, his support for her in the election, and finally in the role he was asked to take on in the fall of 2013: assisting with the government's work on developing the LNG industry. In order to remain as impartial as possible, I will let the reporters tell the story in this section about Gordon's appointment.

It is also worth restating that even Premier Christy Clark did not get to read or influence the content of the manuscript. Surely, if the premier cannot have advance input, then neither can Gordon.

On October 30, 2013, Premier Clark announced her appointment of Gordon Wilson as the LNG Buy BC Advocate. Vaughn Palmer wrote:

> The B.C. Liberals must have been a bit sheepish about this week's appointment of Christy Clark backer Gordon Wilson as advocate on liquefied natural gas given the way they announced it to the public.
>
> Wilson, yesterday's man even among those favoured by Clark's patronage, was not even accorded the dignity of a press release. Rather, the news that he'd been appointed to a $50,000 posting as "liquefied natural gas Buy B.C. advocate" for a starting term of four months came out via the weekly release of cabinet orders.[1]

In fact, Wilson generated quite a bit of ink in the media and online, much of it negative. On November 5, 2013, Les Leyne wrote a column for the Victoria *Times Colonist* entitled "Wilson's Appointment a Bit Mystifying," in which he noted that "Wilson's appointment is utterly baffling to most British Columbians." Leyne went on to point out that Wilson's time working with both major political parties brought a different perspective:

> So with close experience in both camps, he gave Clark the nod last May and she gave him the nod right back. "One of the greatest salesmen I have ever met," she said last week, defending the choice. And with a firm grasp of policy, to boot. "He gets it."[2]

Former CKNW talk show host Rafe Mair, in his online blog, wrote the most scathing post, excerpted below:

Christy Clark is paying Gordon Wilson $12,500 a month for four months—probably a permanent gig if he keeps his nose brown enough. Wilson is going to be an advocate for Liquefied Natural Gas (LNG).

In April last, before this former BC Liberal leader and NDP cabinet minister endorsed Liberal Christy Clark in the May election, Wilson had this to say about LNG:

"The most compelling reason to be concerned about relying on this golden goose is the fact that the markets we are told will buy all we can supply may not materialize as we think, and even if they do, the price they are prepared to pay for our product may be well below what is anticipated."

"The impact of an expanded hydrocarbon economy will certainly speed up global warming and cause us to build a dependency on a revenue stream that originates from processes that are poisoning our atmosphere.

So what happened to Mr. Wilson? Does he have some contract in his pocket for LNG sales from BC to an Asian customer? Has there been some host of angels descend from Heaven, urging Mr. Wilson to get on the side of God and Christy Clark?

Or is he just a grubby political whore whose price is $50,000 a quarter?

After a number of other observations, Mr. Mair concludes that Gordon Wilson "is, price tag stamped on his forehead, a political whore.[3]

Not all media were quite so harsh. Brent Stafford wrote an editorial for Vancouver's *24 Hours*, in which he observed:

It's hard to argue Wilson isn't qualified. His job is to help connect local businesses with companies sourcing goods and services for the LNG industry. His tenure as an NDP minister in such areas as finance should bring special perspective and capability to the execution of his duties.

I also believe Wilson's prior skepticism of the premier's LNG plan, and his concern over the impact on the environment, are assets. A healthy dose of skepticism can reveal new opportunities and by tapping Wilson for this job it certainly shows the premier is confident in her plan.[4]

Why did Wilson take the job after years of avoiding political engagement? In short, he was motivated by Clark's leadership. I will at least allow Gordon the last word on this topic.

I said in my video during the election that we need to support her, and once she is elected we have to roll up our sleeves and help her succeed. I believe she has the potential to be a leader who can build this province into a new social and economic paradigm, and I'm excited by that because that is what I tried to do in 1991 with only limited success. She might succeed.

Honestly, what I wanted to do was fix the BC Ferry system. In retrospect, fixing the BC Ferry system was relatively small, in comparison with working with a multibillion dollar industry that will fundamentally change the economy of BC in a way that will allow us to deal with BC Ferries and education and health care.

When it was first proposed that I take on the LNG Buy BC Program, I was intrigued because I wasn't sure what we were doing.

"Given the vagaries of the international marketplace, it is difficult to put all the eggs in one basket and achieve what you are setting out to do. It is proving extremely difficult to achieve the goal, which is to have three, possibly four, operational LNG plants in BC and become a significant player in global LNG supply.

"The premier has made the statement that the BC LNG industry will be the cleanest in the world, and that has some significant recommendations to agencies of government that have to make sure it is correct: Oil and Gas Commission, with strict oversight on drilling, fracking, capture, piping, and on cryogenic process; Ministry of Environment, with strict oversight on GHG emissions at the wellhead, along pipeline routes, at the cryogenic plants and the fuelling stations; the

geotechnical data required to monitor areas where potential seismic
activity can occur is stringent.

Christy's motivation is to build an economy that generates wealth for
the province so that we don't have to borrow from our left pocket to
pay our right pocket. Generate wealth so you have the ability to spend.
That's why she's embraced the LNG industry, so that not only will we
have wealth generated in BC, but we can contribute to the reduction
of greenhouse gases in places like China. She understands you need a
stable investment climate to bring the investments in.[5]

Financial security, once seen as a middling value, is now students' top goal. In 1966, students felt it was important to at least present themselves as philosophical and meaning-driven people. By 1990, they no longer felt the need to present themselves that way. They felt it perfectly acceptable to say they were primarily interested in money.

DAVID BROOKS

CHAPTER 26

CLIMATE CHANGE

PREMIER CLARK IS of the opinion that no one can credibly argue against climate change anymore. "Look around: the pine beetle, forest fires, visible changes in nature. We have to react to these issues as a government, and they make it very real. Rather than a scientific argument, climate change became a very real argument. When the red jellyfish started to show up in the Gulf Islands, when some songbirds disappeared; these changes over time became quite evident. I noticed the change in the ocean that I have grown up swimming in; the complete absence of starfish in some areas."

Within BC, whether in conventional media or social media, Premier Christy Clark is often portrayed by her detractors as pro-corporation and pro-resource extraction and as one who pursues these ends to the detriment of community interest, workers, and especially, the environment. Outside BC, however, Premier Christy Clark is often called upon to tell the province's success story for its world leadership on the carbon tax and advancement of green energy. There is a wide gap in the messaging. Is it simply a case of the prophet not being recognized in her hometown?

"Climate change is a central concern for my government," Clark says. "The world is getting warmer, and it is going to make parts of the earth uninhabitable. It is going to create scarcity, especially in terms of water and food supply."

I had an opportunity to interview the premier shortly after she returned from the World Climate Summit in Paris in the fall of 2015, where Canada was represented by several provinces, including BC, and by our newly elected prime minister, Justin Trudeau. By all accounts, Canada provided a welcome contribution, and in the national media coverage, I noticed how often Premier Clark was interviewed about BC's success in battling climate change through legislative efforts. It felt like quite a contrast from what I saw and heard when Premier Clark was at home; I heard more about NDP premier Rachel Notley's efforts in Alberta.

Clark sees local food production as intrinsic to the planning for a changing climate. Although we have not seen much written about Clark's government's support of local food, she is quite passionate about it. "In my first budget as premier, we excused agricultural production from carbon tax including greenhouse production. We knew it was more important to support the food producers and give them every opportunity to succeed, than it was to include them. Besides, there is a huge connection between local food production and climate change."

One of Christy Clark's government's early initiatives to support a connection between local food producers and local buyers was the resurrection of the Farmers' Market Nutrition Coupon Program. This was a pilot project from 2007 to 2010, when it was cancelled.[1] It was restarted as a full project in 2012 with a $2 million dollar investment, expanded in 2013 with an additional $2 million investment, and provided $750,000 in 2014 by the government to sustain the program through 2015.

In 2015, forty-eight communities participated through their local farmers' markets. There are now additional sponsors, and Premier Clark is very proud of the outcome. This program provides lower-income families, pregnant women, and seniors with coupons for locally produced food. Today, the program is run by health minister Terry Lake.

The Farmers' Market Nutrition Coupon Program runs from July through October and is administered by the BC Association of Farmers'

Markets. According to the website, "These coupons can be spent at all BC farmers' markets that participate in the FMNCP to purchase fruits, vegetables, cheese, eggs, nuts, fish and meat. Each household enrolled in the program is eligible to receive a minimum of $15/week in coupons."[2]

When the program was expanded, representatives from some agencies expressed its benefit. "We are so pleased to be participating in the Farmers' Market Nutrition Coupon Program," says Suzan Goguen, executive director with the Seniors Outreach Society. "Each week, our seniors are able to use the coupons to visit their local market, where they can access fresh, nutritious food and engage with their community—this helps their overall health and well-being tremendously."

To participate in the program, farmers' markets partner with a community agency that works to provide nutrition, cooking, or healthy lifestyle skills-building programs to lower-income British Columbians. Participants then gain the skills and knowledge needed to help them eat healthier and make the most of their local farmers' markets.[3]

The premier emphasizes the focus. "You want to make it easier and more competitive for people to produce food locally. It's fresher, better for you, and you don't have to travel to get it."

After the success of these and other programs, the BC government expanded its emphasis on agriculture in the 2016 Throne Speech, announcing:

Last year, at $3 billion, was the highest ever sales of B.C. food and beverage products. This year, your government will continue its work to increase provincial revenues in agrifoods and seafoods to $15 billion a year by 2020.

Climate change and increasing demands on water are challenging global agricultural production, in particular in the US and Mexico where much of our fresh produce is grown. Combined with the current low Canadian dollar, this creates rising food prices, which are putting a strain on B.C. families.

This year, your government will build on those successes by increasing its financial support for the [Agricultural Land]

Commission and moving forward with a tax credit for farmers that donate food to non-profits. And in November, the first ever provincial agrifoods conference will be held in Kelowna, focusing on food supply security for B.C.

Your government will expand on these efforts by piloting work with industry, local governments, and community organizations to encourage and support British Columbians to Buy Local, Grow Local. This work will get more British Columbians engaged in growing food at home and in their communities. It will provide another source of fresh fruits and vegetables, and further strengthen the connections between British Columbians, our communities, and our agricultural sector.

ONE TANGIBLE CLIMATE change problem the Clark government must face every year is BC's growing wildfire season. This is particularly relevant to Clark now that she represents a riding in Kelowna, which faces the summer desert conditions of the Okanagan Valley.

"When it comes to wildfires," she says, "the trend is that they are getting worse. It is not always more expensive, because we are getting better at fighting them. The weather is more unpredictable, the earth is getting hotter, the trees are getting less resilient because of the pine beetle infestation."

Clark is referring to the infestation of the mountain pine beetle, spreading in part because winters are not as cold as they were, so the beetle survives winters. The pine beetle is killing large tracts of Interior forests. "The epidemic peaked in 2005: total cumulative losses from the outbreak are projected to be 752 million cubic metres (58 percent) of the merchantable pine volume by 2017."[4]

In the summer of 2015, the fires were visible from Clark's deck in her riding. Clark says, "Having a home in West Kelowna and so many more friends who are directly affected by wildfires absolutely changes your view."

Clark was in attendance, meeting with the people fighting the fires, making sure they had everything they needed and that they knew the government support was there for them. "Steve Thomson [Minister of Forests] deserves a lot of credit for making that fast response happen. He didn't even take a vacation that summer."

There was a loud and large fight with Premier Clark in the summer of 2015 over her government's reluctance to use the Mars Water Bombers for the fires. I asked her about this. "Experts we rely on to fight fires say there is a very limited use for the Mars Water Bomber. There are some circumstances where we put it to use."

She is reluctant to accept the critics' view that the Mars Water Bombers are better technology. "If you had a heart attack and you needed help, would you want the ambulance to be a new one, or one designed before the Second World War? When you are located by one of the fires, you want the best fire crew and the best fire equipment, and you want to know that everything is managed right away."

The centrepiece of the Clark government's commitment to battle climate change is the carbon tax, first introduced by the Campbell government. This tax, in its current form, is widely seen as a model for other jurisdictions.

Says Clark, "The unique element that BC brings to the international discussion is that we have been able to both grow the economy and reduce our impact on the environment at the same time. There is almost nowhere in the world that can make the same claim."

She has heard many citing the plans of Premier Rachel Notley's government in Alberta but points out that "it is going to be 2018 before Alberta even *starts* to catch up with us."

Clark has been sought out by leaders asking how to succeed with a carbon tax. "It is why so many global players are so excited about what we have done, and we saw that when we attended the Paris Climate conference in November of 2015 as part of the Canadian delegation.

"The question for any leader when dealing with all the responsibilities and the reality of climate change is: 'How much can we do on our own without harming our economy, so the businesses that employ people in the most environmentally sound framework do not leave for a neighbouring jurisdiction, where they can pollute without restrictions?'"

Because BC introduced a carbon tax early, there was a risk that investment would go to other jurisdictions to avoid the tax. So far, that has not happened. When I asked the premier about the Opposition's criticisms that her government isn't doing enough, she quickly shot back.

"I would ask, where is the NDP credibility on this? They ran an election against the carbon tax. Carbon pricing," she points out, "is an essential element of controlling greenhouse gases. This is the consensus globally."

She does not believe the solution to global agreement on combatting climate change can be achieved overnight. "My experience with these conferences is that progress is incremental. You can't have this kind of progress unless you come together and you have the international leadership pushing the agenda forward. The meeting itself forces everyone to show up with a plan. It isn't like a peace negotiation; we are not coming up with the Marshall Plan." She is quite animated about this. "It's a process."

A memorable moment in Paris at the climate conference was her interaction with the youth delegates for BC. The tone of the meeting was not particularly friendly because the youth were adamant that more action had to be taken immediately.

"Youth delegates were very passionate, and one of them was arguing to me that BC should not do LNG because it will mean an increase in our emissions and we need to be solely focused on reducing BC's 0.2 percent contribution to the world's emissions."

The premier was referring to the fact that BC contributes only 0.2 percent of the world's carbon emissions, and was putting the LNG addition in this context. Her reply attempted to paint the bigger picture. "That is a noble plan, but when we all share the air, how do we get there? If we shut down our LNG plan, we lose the chance to stop some of the 150 coal plants in China to be built."

Those same coal plants in China were cited at the conference as some of the *worst potential future contributors* to global carbon emissions. The premier did not make much impact on the young idealist, however. "His view was that everything was possible, right now." She clearly respected his opinion even as she disagreed with it. "It does make you feel bad, to say to such an idealistic young person that, yes, we can get there, but we have to have a plan, and it takes some work."

Unlike the youth delegates, the premier has to try to work with everyone. "Government's job is to figure out how to get to the outcome."

When asked how the government can be proactive in dealing with climate change, rather than simply dealing with reducing carbon output, she says, "We have to focus on changing our built environment, to make our houses, commercial, and office buildings as energy efficient as humanly possible. We need to develop a strategy around passive buildings, combine that with rapid transit, ensure efficient transmission of power, so we don't lose energy as we transmit it."

She is not keen to raise the carbon tax immediately, and has asked the federal government not to interfere with BC's structure. She said the Paris conference was useful. "[The] climate change panel came up with some very interesting, challenging ideas."

BC Business conducted a poll in December of 2015, after the Paris conference, and asked the business community if Christy Clark should raise the carbon tax. Surprisingly, over 85 percent of respondents said yes; however, most want the premier to wait until 2018, when the carbon-tax freeze is scheduled to end.

Premier Clark understands that raising taxes sometimes has consequences that negatively affect jobs, and she sees her mandate to include job protection, and in fact job *creation*, while she protects the environment. "If it not done carefully, you will bring economic growth to a full stop. Polluters will just move and go somewhere else."

She believes that only a government that is 'business friendly' can succeed in implementing carbon taxing by working with industry. "Carbon pricing that works is almost always brought in by governments that understand how the economy works."

Prior to Clark's attendance at the Paris climate conference, there was pressure on her to abandon her government's commitment to developing LNG. Travis Lupick of the *Georgia Straight* writes:

> On November 30, B.C. premier Christy Clark will accompany Prime Minister Justin Trudeau to France for the United Nations conference on climate change.
>
> The meeting—convened with the ambitious goal of drafting a legally binding agreement on greenhouse-gas emissions—will see Clark take something of a leadership role on the environment.
>
> But a number of organizations—such as the City of Vancouver, UBC, and Concert Properties—are calling on Clark to do more to prove B.C. deserves that position of leadership.
>
> Karen Tam Wu, program director of building and urban solutions with the Pembina Institute, told the *Straight* there is a disconnect between the province's green image and the premier's enthusiasm for liquefied natural gas (LNG).
>
> "To be blunt, the way that B.C. is planning to develop LNG—to their ideal of three LNG terminals—we are going to have a really hard time meeting our climate targets," she said.
>
> Tam Wu said there is time for B.C. to change its policies on natural gas, noting the province is still drafting a "climate leadership plan." That document, which enters a second round of public consultation this December, will outline how B.C. intends to reduce carbon emissions to 33 percent below 2007 levels by 2020 and 80 percent below 2007 levels by 2050.
>
> "What we really need to do is look at how we, as a province, are moving off of fossil fuels, not developing new fossil fuels," she said.[5]

The election of the Justin Trudeau Liberal government in Ottawa in October 2015 changed the dynamic between BC and Ottawa on many topics, and climate change was high on the list. Canada moved from a

Conservative government that suppressed discussion and scientific evidence on climate change to a Liberal government that appointed a Minister of Environment and Climate Change.

Premier Clark now had the ability to work with the federal government on a national strategy on climate change that allowed BC to work in conjunction with the other provinces, rather than continue to set its tax and emissions rates on its own. As Justine Hunter reported in the *Globe and Mail*:

> British Columbia is poised to abandon its legislated targets for reducing greenhouse-gas emissions, but the decision will be shaped by how the rest of the country steps up to the climate challenge set by Prime Minister Justin Trudeau.
>
> Premier Christy Clark said in an interview Wednesday the B.C. targets for 2020 will be extremely difficult to meet... However, the Premier said soon-to-be-developed national targets will determine whether she is forced to scrap or amend British Columbia's climate-change law.
>
> Ms. Clark said she is willing to make further changes to drive larger reductions in the province's greenhouse-gas (GHG) output, but she said British Columbia has already done more than its share in the national context. "We have done better than anyone else in the country. Will the new targets recognize all the work we have been doing for eight years while almost nobody else has been doing anything? We don't have coal plants to shut down. There is no low-hanging fruit left in British Columbia."
>
> Environmentalists have argued that the Premier's ambitions to launch a liquefied natural gas (LNG) industry in British Columbia would derail the climate targets that were set in 2007 under then-premier Gordon Campbell.
>
> The government hopes to see five LNG plants built on the coast, which could increase the province's annual emissions by 13 million tonnes—if industry meets the benchmarks set by the province.

The province is eager to talk to Ottawa about the federal Liberal commitment to spend $20-billion on green infrastructure over five years, and to create a $2-billion "low-carbon economy trust" to fund projects that reduce carbon emissions. Those dollars could help British Columbia move toward its climate targets—the old ones, or more likely, new ones.[6]

At the end of 2015, press gallery reporter Tom Fletcher did an interview with Premier Clark and captured some of her thoughts on LNG and climate change, excerpted here:

TF: I want to start with your trip to the UN climate conference in Paris. Did you speak about natural gas as a transition fuel, and did you find support for that idea?

PCC: Yes and yes. The new government in Ottawa is a big supporter of our LNG plan, and part of the reason for that is that they also see it as a way forward for Canada to make a huge contribution to fighting global climate change.

There are 150 coal plants on the books to be built in China today. The only way that those plants and the ones that come after will be stopped is if they have a transitional fuel to move to. We all want to get to 100 percent renewables one of these days, but that's not going to happen in the next few years, probably not for the next many years.

TF: Prime Minister Trudeau campaigned against subsidies for fossil fuels. That was interpreted by some people as cancelling capital cost allowances for an LNG plant, passed by the previous Stephen Harper government.

PCC: They continue to support that change made by the previous government. They have publicly endorsed that position. LNG will be a source of emissions for Canada, but overall it's going to be a big favour to the world.

When asked why it was that BC was likely not going to hit its 2020 emission target, legislated back in 2007, Clark explained that the lack of movement elsewhere in the country meant that British Columbia needed to pull back. If too big a regulatory gap developed between BC and other provinces, then it could cost British Columbians jobs or damage the economy. In addition, she said:

> PCC: What I saw this year was developing countries, especially China, making a firm commitment to reduce their emissions. They say they're going to peak in 2030. The only way for them to do that is to move to a greater degree to natural gas, and the bulk of their industry is still located on the east coast of their country, a long way from Russia and close to B.C.
>
> So I think the concern about climate change is going to rebalance the market for natural gas . . . I do know that countries are going to be looking to natural gas as the primary solution to the climate change issues they're trying to resolve.
>
> TF: Veresen has just green-lighted a second gas processing plant for the Montney region, so they must have some confidence.
>
> PCC: Producers have invested $20 billion in B.C. so far and there's no sign of that slowing down. These are long, 30-year agreements. I'm going to be 80 by the time some of those agreements expire. Goldman Sachs couldn't predict this latest downturn in oil. I don't think anybody knows what it's going to look like in 30 years.[7]

In January 2016, Premier Clark called for further public input on the climate change strategy for BC. In a story published on Kelowna's main news website, Darren Handschuh reported that the province wants to know what everyday citizens have to say about climate change, with excerpts here:

> Premier Christy Clark is inviting families and members of the public to tell government what they would like to see in B.C.'s new Climate Leadership Plan.
>
> "The input British Columbians have provided to date has been very helpful as we develop B.C.'s new Climate Leadership Plan," said

Clark. "Now is the time to engage even further in the conversation, as we seek to build on our global leadership through our next set of climate actions."

In addition, Clark also announced she will chair a new cabinet working group on climate leadership that will oversee government's climate actions and policies, while considering current and future provincial climate action goals.

In the summer, the province launched the first round of consultation towards a new plan, engaging with the public and First Nations, as well as local governments and organizations to find out their climate action priorities. Government also received more than 200 written submissions and more than 300 template letters.

This feedback helped inform the Climate Leadership Team as they created their recommendations for the new climate plan, which were submitted in late November.

The province's final Climate Leadership Plan will be released this spring, and will include new climate actions to drive down emissions while supporting a growing economy.[8]

Premier Clark often speaks of her relationship with her son, Hamish, and how he encounters the world. For Hamish, climate change and the natural world is more direct.

"If we were discussing the heart of the problem with climate change, it would be Hamish describing it to me. When he is outside in nature, his response is to the beauty of it, and he wants to preserve it. What Hamish would notice the most is the impact on polar bears and penguins, that there are fewer of them and they don't have adequate food anymore."[9]

She understands that it is her job to contemplate and implement policy, to work toward the best future for her son and future generations.

There's joy in a life filled with interdependence with others, in a life filled with gratitude, reverence, and admiration. There's joy in freely chosen obedience to people, ideas, and commitments greater than oneself... There's an aesthetic joy we feel in morally good action, which makes all other joys seem paltry and easy to forsake.

DAVID BROOKS

CHAPTER 27

WATER, WATER EVERYWHERE

PREMIER CLARK IS aware that climate change will create water and food scarcity, and she is aware that water is a precious resource in BC. "We cannot keep treating water like it is an endless resource," she says. "We treated fish that way, we treated trees that way, we have treated all kinds of different resources that way."

Her government undertook a rewriting of the *Water Act* as part of the move to protect the resource through regulation, and as with so many issues in BC, this initiative became a hot topic.

"This will be the first time we have ever had regulated water use in the province; we are updating a one hundred–year-old act." You can tell that this is something of which she is proud. "We need a framework to regulate water use. There was no framework *at all* to regulate ground-water use."

As she shares her passion for protecting our water resource with me, I had to laugh, and it interrupted the interview.

"You realize the irony in the difference between what you are telling me," I asked, "and what is being discussed publicly, right?"

"Absolutely, and I honestly don't know what to do about it. We are just going to stay focused on protecting our water, and hope the public figures that out."

In early 2015, stories started to appear claiming that the BC government was giving away our water resources and linking this to multinational corporate giant Nestlé, which has a bottling plant in BC drawing from groundwater. Prominent New Democrat Bill Tieleman wrote an article for the *Tyee* that helped start the momentum.

> Have you ever paid $2.25 for a bottle of water? Of course, and you can pay a lot more than that if you go to a Vancouver Canucks game, a concert, movie theatre or restaurant.
>
> So what if you could pay $2.25 not for a 500-millilitre bottle, not for a big office cooler full, but for 1 million litres of water?
>
> Sounds ridiculous given the retail price, but that's the unbelievably low rate the BC Liberal government has given to giant multinational firm Nestlé and others to extract fresh, clean groundwater to bottle and sell for exorbitant profits.
>
> The price is so outrageous I have to repeat it. Nestlé Waters Canada pays the province just $2.25 for every million litres of water.
>
> The total estimated price of all the water Nestlé will bottle in B.C. over an entire year is—wait for it—$562 a year!
>
> That's an improvement, if you can believe it, because until recently they got it all for free.[1]

The article goes on in the same outraged tone, ending with the suggestion that Nestlé and other water users be metred and forced to pay more to buy our groundwater.

This article was widely circulated on social-media, with many negative comments on the government's policy, including that this was another indication of how the Clark government was connected to large corporations. I know this because I spend time in the social-media universe, and with over four thousand on my news feed, can pick up on regional trending.

Shortly after Tieleman's article appeared, furious environmentalists started an online petition by SumofUs. I was seeing the petition scroll by regularly, and was paying a bit of attention.

Now, a bit of personal context. One of the issues I spent the most time on when I was elected in the early nineties was water and the protection of our water resources from commoditization. I was outspoken on it then and have followed the issue since. I started an environmental organization in 2008 called Pebble in the Pond Environmental Society, recruiting some other environmental activist friends, and we lobbied for a number of environmental concerns; initially, one of our targets was bottled water. In 2013, the mandate expanded to include water quantity and quality protection.

On social media, I have been fairly well connected with environmental activists because I am one of them. I was heavily criticized by many of them for supporting Clark in the 2013 election; however, I am not a particularly partisan person, so generally they had left me alone in my political support of her government. I have been a supporter of SumOfUs for many years and follow their campaigns. Most of them are excellent.

I saw the SumOfUs petition scroll by, supported by some of my Facebook friends, with the headline "Nestlé Is About to Suck BC Dry," and the petition text called on the government to "Charge a fair price for Canada's groundwater! Commit now to review the water rates!"[2] The petition was popular, and on March 6, Dan Fumano of the *Province* reported:

> More than 82,000 people have signed a petition against the government's plans to sell B.C.'s water for $2.25 per million litres.
>
> "It is outrageous," says the online petition from SumOfUs.org, that corporations can buy water "for next to nothing."
>
> B.C.'s Water Sustainability Act (WSA), which comes into effect next January and replaces the province's century-old water legislation, has been heralded as a major step forward.
>
> Last month, the government unveiled the new water pricing structure, which will include, for the first time in B.C.'s history, groundwater being regulated and subject to fees and rentals.
>
> Critics said that, while it's a step in the right direction, the prices are still not close to capturing the resource's value ... water rates for

industrial users, which are a fraction of what some provinces charge, are "like a giveaway" to corporations, critics say.

NDP environment critic Spencer Chandra Herbert said the new legislation is "promising," but questioned whether it would actually live up to its promise, or just remain "nice words on paper."

"A lot of business groups, community groups, farmers—they want to see better protection for their water. I'm just worried we're not going to get it."

When Chandra Herbert raised the issue last month in the legislature, Environment Minister Mary Polak replied that British Columbians are "quite proud" that B.C. "has never engaged in the selling of water as a commodity."

Polak said, "We don't sell water. We charge administration fees for the management of that resource."[3]

Unfortunately, Polak's message, tucked away as it was at the end of the story, was lost on the critics, perhaps because they did not understand how important it was to distinguish between an administrative fee and a per-litre sales price. The water issue and petition made the rounds of social media throughout the spring of 2015, and when BC experienced an early summer, a lack of rain, and many water shortages, the petition's momentum escalated.

Summer arrived in May that year. It was very hot and dry, and by July, many communities were on stage four water restrictions, with no sign of rain. People were very angry because they believed the government was giving away our precious resource to greedy corporations and ignoring the needs of the people.

Radio programs started to expand discussion about water beyond the headlines, and callers were very worried about groundwater drying up as we continued to give it away. On CBC's *On the Coast*, Environment Minister Mary Polak was interviewed on the issue, as were the critics of her government's plans. Polak remained focused and articulate and on message; petition organizer Liz

McDowell was equally articulate, and was angrier and more resolved to have Polak change her water policy and charge money to sell our water.

Criticism against Nestlé is picking up momentum once again as drought plagues B.C., wildfires rage in parts of the south coast, and residents are facing water restrictions.

Nestlé Waters Canada bottles roughly 265 million litres of water from B.C. every year.

Starting in 2016, it will have to pay $2.25 per million litres due to new regulations.

"It's simply scandalous that a company like Nestlé can take hundreds of millions of litres of groundwater at basically pennies at the same time as other B.C. residents are being asked to conserve water because it's in the middle of a drought," said Liz McDowell, campaign director of an online petition opposing the new regulations.

Nestlé Waters Canada defends its water use, noting it is not withdrawing water from rivers, lakes and streams that are currently affected by drought. Instead, the company draws its water from a groundwater aquifer, said corporate affairs director John Challinor.

"We withdraw less than one percent of the available groundwater in the Kawkawa Lake sub-watershed," he said in a written statement.

The Kawkawa Lake sub-watershed is near the District of Hope, where stage four water restrictions are prohibiting outdoor water activities and watering of lawns.

McDowell says the $2.25-rate is much lower than those set in other parts of Canada. Her petition, which began earlier in the year and has now gained more than 160,000 signatures, is urging the B.C. government to set a much higher water rate to encourage conservation.

"We're fortunate we do have groundwater reserves at the moment, but in California, after two, three, four drought years in a row, their groundwater reserves are actually running dry," she told *On the Coast's* Stephen Quinn.

"In B.C., right now we're in a position where we can actually stop and think responsibly and make sure we can preserve our groundwater for years to come."

McDowell says she will present the petition to Environment Minister Mary Polak once it reaches 200,000 signatures.

In an interview with Stephen Quinn on the *Early Edition*, Polak insisted that the province is not selling water but just charging a fee to industry for "accessing" the water for free.

She says raising fees could raise legal questions about who owns the water in the province.

"That's a dangerous thing in these days when water is fought over around the world and we see what's happening in California," she says. "We will never sell that right of ownership. We will allow access but it is tightly controlled."

"If we want to talk about preserving water, it's about making sure we have strong laws to regulate how much anyone can use—whether they're Nestlé or anyone else—and ensuring that we never open to allowing multinational corporations to come in to British Columbia and actually buy ownership of some water.[4]

By July 2015, I had just started work on this book, and one of the first people I interviewed was the premier's brother, Bruce Clark. I conducted the interview from my office, which is attached to the barn on our sheep farm, and he was being really irritating as I interviewed him because he kept asking me questions about what I would do differently if I were in the government.

I kept telling him that it didn't matter what I had to say about anything because I was in no position to do anything, but he kept returning to his questions, in between discussions about other issues.

Finally, exasperated, I went into a rant.

"Okay," I said, "well, if I planned to say anything about anything right now, I would go after the people promoting that darn water petition and tell them that they are wrong. If the government actually says yes to them, we will lose control over water!"

My rant was actually a bit longer than that because I slipped into my policy-wonk mode.

When I was done, Bruce thought for moment and then said, "You should come out publicly and say that. I forgot about NAFTA, and I think most people have forgotten about that."

"Bruce, no one has *forgotten* about NAFTA. And anyway I live on a sheep farm in the middle of nowhere. Who cares what I have to say? Even if I did say something, where would I say it?"

I laughed, and tried to move on.

"Say something," he insisted.

Bruce is not on social media, so I really have no idea where he expected me to say something, but my only option would have been on my Facebook page.

The next day, I received another email from someone trying to get me to sign the petition to save BC's water. I forwarded it to Bruce with a one-word note: "Momentum."

He wrote back: "Say something."

It was a Saturday, and although normally I would have been working outside on the farm, I was in my office setting up some book interviews. I decided that I would take five minutes and post something on my Facebook page. This is what I posted:

I have to make a statement about the growing hysteria around BC's laws on water, which terrifies me because the campaign is wrong. Do NOT demand that the province charge Nestlé money, unless you want to open the door to massive water sales in BC. Please read my full statement very carefully because water is more important than politics. Context: I have been campaigning against bulk water exports since the late 1980s, arguing against the Free Trade Agreement, and later NAFTA, which defined water as a commodity eligible for export, and contained clauses locking us into sustained export levels regardless of our domestic need. Recent news stories alleging that the BC government is "giving away" our water to Nestlé and stating that the government should be "charging a fair price" for it are dangerous. Currently, Nestlé pays the same fees that everyone else pays for access to water. Nestlé is on the record saying they will pay a fair price and in fact want the province to do an inventory of its fresh water. Understand this as you sign the petition demanding that the province charge Nestlé for water: you are lobbying our government to turn our water into a commodity for sale. That's what you are doing when you post articles and petitions. You will make Nestlé very happy if you succeed, because then we can never turn off the taps due to the international trade deals in place.

Do not sign the petition; do not ask that our water become a commodity. When Environment Minister Mary Polak says, "We don't sell water. We charge administration fees for the management of that resource," she is trying to tell us this—we are not selling it. And that is the only way we can protect it. By not selling it. Honestly, this campaign terrifies me. Every headline I have read is misleading and intended to make you angry and lobby for something very bad. Please share.[5]

I shut my computer down and went outside to play farming. By the time I was halfway across the field, my phone started to act funny. It was sending me notifications of shares and likes of my post, and they didn't make sense. It was only the next day, when I logged on to my computer to interview Keith Baldrey for this book that I realized the action on the post was happening so fast, my phone couldn't keep up with it. The post had over five thousand shares in less than twenty-four hours, and while Keith and I were talking, it went up by several hundred more.

My phone started ringing with media, and I started to receive emails and private messages asking for interviews. As of January 16, 2015, this post had over 19,000 direct shares, 5,664 likes, and over 730 comments. It had also been translated into other languages by unknown people, shared online in those languages, it had been turned into a jpeg (photo) and posted to other social media, and it had been circulated as a link on websites and Twitter.

I didn't expect any of this response and found myself caught up in the debate, which spread to Quebec, where they were dealing with a similar issue. I had to keep reminding the media that I did not speak on behalf of anyone. At any rate, I was happy to have provided a different perspective, and I did blame Bruce Clark for all of it.

When I told him, he said, "I'm happy to be blamed for something good for a change."

My post made me a target, as well, for those who were keen on seeing the petition succeed. I lost a few friends during this, as the NDP circulated stories claiming that I was motivated to advocate for the government because my husband had been appointed to assist the LNG industry.

At any rate, given the extreme negativity against the Clark government for its stated indifference to protecting the water resource, Premier Clark's passion and clarity on the issue when I interviewed her were a great contrast.

Clark spoke of her government's vision around water. "Water is the most essential substance on earth. If we are going to regulate something, surely that should be the first, not the last. It seems to make sense that this resource, essential to human life, would also be regulated to ensure that we always have enough to share, and that it is protected from pollution."

She notes that the NDP had been loud the year before the petition about protecting water. "The complaints came from the same people who, last year, were arguing that we should never make water into a commodity. Last year they were saying you can never put a price on water, and this year they were saying you have to put a price on water, but just for some people."

Clark knows you have to treat all companies the same, you cannot pick and choose who you charge. "What about all the craft breweries in Vancouver? They use tons of water. Do we want to shut them down? The biggest users of water are some of the biggest job producers, plus the agriculture sector relies on it."

Her government is clear on its water policy. "We do not support bulk water exports. Water is only going to become a more valuable resource; what a blessing it will be to be a British Columbian and have this abundance of clean water."

Scarcity, she knows, will be a growing concern. "As California gets drier and drier, their demand for our water is only going to grow. We have to think fifty years ahead; we have to be able to resist that demand. We do not want to be in a position where we have to send our water south with no choice."

When the SumOfUs petition reached 200,000 signatures, it was presented to the Clark government, and Environment Minister Mary Polak committed to review the issue of fees.

Joy is not produced because others praise you. Joy emanates unbidden and unforced. Joy comes as a gift when you least expect it. At those fleeting moments you know why you were put here and what truth you serve . . . you will feel a satisfaction, a silence, a peace—a hush. Those moments are blessings and the signs of a beautiful life.

DAVID BROOKS

CHAPTER 28

A FULFILLING JOURNEY

WROTE THIS BOOK because I felt that there was a story that needed to be told about Premier Christy Clark, because I felt the portrayal of her as a heartless, arrogant, corporate sellout did not match the truth about her or the work she was doing on our behalf. If we are going to judge a government, we need to know what it is *actually doing* first. British Columbia has a brutal political climate, with an entrenched partisan dialogue that often distorts our view of leaders and their accomplishments.

Although this book cannot address everything, I am going to touch briefly on a few key topics often raised about Christy Clark. First is the sale of liquor in grocery stores, part of the push to modernize liquor laws. Any talk of changes in the past was controversial, because government liquor stores provide good union jobs, and privatization threatens them. Premier Christy Clark compromised: British Columbians' wishes for easier access would be met by allowing only BC-made alcohol in grocery stores. As of February 2016, a few grocery stores were blazing a new trail, with one store boasting *800* BC-made wine products.

Just how bizarre are some headlines about BC? In the *National Post*, Brian Hutchinson reports, "group sex, illegal pot get ok in vancouver, but wine in grocery stores? no way." The Vancouver Health Authority made doomsday predictions to Vancouver City Council in its staunch opposition to wine and beer in grocery stores:

> "Its recreational use comes at a price," the VCH submission reads. "Alcohol causes an estimated 7.1 per cent of all premature deaths in Canada and 9.3 per cent of premature death and disability combined ..."
>
> This from a public health agency that endorses group-sex parties inside local establishments, where alcohol is served. One Vancouver business offers regular "no-holes-barred-sex" parties with VCH representatives on hand to dispense condoms and pamphlets. No issues there.
>
> Vancouver is also home to dozens of illegal marijuana dispensaries ... the city has decided to regulate cannabis sales, and collect from its pot shops exorbitant business licence fees.
>
> Given all that, it seems strange that Vancouver city council would reject one of the most popular provincial government initiatives in years.[1]

Vancouver Mayor Gregor Robertson is a former NDP MLA. His council is mainly Vision Party members, aligned with the NDP, and although they do not say they are speaking from a partisan perspective, their opposition is hard to understand otherwise.

This should have been seen as a win-win-win: preservation of government store jobs, expansion of the market for BC products, and an answer for the working parents who just want to grab some wine or beer while shopping with their kids. Instead, a number of groups continue to wage a vocal and nasty campaign against Clark's changes.

The Clark government is often accused of being uncaring. When I asked Deputy Premier Rich Coleman, "What do you say to those who say your government doesn't care about the little people?" Coleman replied with a passionate response, "They are wrong. This government has done more for people on the social side than any other government in history."

Speaking in the fall of 2015, Coleman said that the government knew it had to take quick action a few years ago, and they did, and it has worked.

"The number of homeless in the past five years has dropped by six thousand people and twenty thousand families get a cheque every month to assist with their rent."

Coleman explained that the government made a deliberate decision to help families afford to pay for rental housing in the communities where they wanted to live, rather than set up a government agency to build government-owned rentals. "This program means you don't stigmatize people, they are just living in the community in regular housing.

"Anyone making $36,000 per family, or less, can apply for help. To try to build 20,000 units in housing takes time, but allowing people to choose where to live and helping them with their rent is a lot quicker, which is a better outcome for the children." It isn't just families with children who benefit, he said. "There are 18,000 seniors who also receive rent assistance from the province. We now have a rent assistance program for people in shelters. People waiting for affordable rentals have a long wait, there's a long list, so we are focusing on the rent assistance."

Even though in BC this initiative has received little attention, Coleman said, "I've sat in rooms with other ministers responsible for housing across Canada who say that BC has the most successful housing strategy in the country. It has worked very well."

If housing policy may be working, there are few who will agree that the BC Ferry Corporation policies are making coastal communities happy. This is a corporation started in July 1958 by Premier W.A.C. Bennett with two ships, two terminals, and around 200 employees. "50 years later, BC Ferries is one of the largest, most sophisticated ferry systems in the world, with 35 vessels servicing 47 destinations ... The staff now exceeds 4,700 in the summer months."[2]

There was considerable controversy with BC Ferries during the NDP government, and when Gordon Campbell became premier, his government changed the structure of the ferry corporation so that it was incorporated as a private company, with one shareholder: the BC government. This structure continues to be highly controversial, with coastal communities insisting that the ferry corporation is an extension of the highways.

There are many loud calls to return BC Ferries to Crown Corporation status; however, the current debt load of the corporation would, if added to the BC government books, potentially affect the province's credit rating. At this point, government is in a bit of a stalemate with coastal communities demanding immediate action to make the ferry service more affordable and, in some regions, more reliable.

Local governments on Vancouver Island, the Gulf Islands, Powell River, and the Sunshine Coast are working with their MLAs, most of whom are NDP, to try to force change. This issue is too big for the scope of this chapter, but suffice it to say it is reasonable to expect the problems with BC Ferries to be shouted more loudly the closer we get to the 2017 election. Cynical observers may comment that the Liberal government has little at stake, since most of the MLAs in this area are already NDP.

Another hot issue is in forestry, regarding worker safety and private forest lands. I had a front seat in watching one controversy unfold in the summer of 2015 when, on Earth Day, chainsaws started up in Powell River and trees started to fall on land that was privately owned but that the community believed was publicly protected. I spent quite a bit of time with the person who emerged as one of the leaders of the fight to protect the urban forest, Jenny Garden, the granddaughter of former NDP MLA Frank Garden.

"When I was introduced to forestry practices, it was in the fight to protect an urban forest in Powell River. I remember when my jaw dropped and I cried after reading about the *Private Managed Forest Lands Act* (PMFLA) and the Managed Forest Council (MFC). Right there, in black and white, I read about the ineffectiveness of the Council and the meagre fees on companies that break the rules... [T]he Council is nothing more than a bunch of industry buddies sitting around a table who refuse to issue a stop work order."

Jenny Garden, like her grandfather, is a fighter who advocates for her community. She said, "Indeed, not once has a stop-work order been issued by the Managed Forest Council."

In short, private forest companies that voluntarily agree to meet a higher standard of forestry register to receive a discount in provincial taxes. It is reasonable to expect that, if they violate the agreement, they receive huge fines or have to pay back the discount they received in their tax bill. This is not the case, as we learned first-hand in 2015.

Garden said, "It's the fox guarding the henhouse, and it's left all the chickens vulnerable to being devoured. It's horrible . . . tax breaks for registering with an organization with zero enforcement track record."

Another leading voice speaking out against forest practices was seasoned forestry worker and danger-tree assessor, Jason Down. With a market-driven focus, Down said the private land system needs more regulation, not less. "Ecosystems and economies do not recognise lines on a map separating PMFL (private) from TFL (public) or Crown Land."

In Powell River, the creek systems and adjacent riparian zones run through protected parks, then the private lands, and then into the ocean. There were at least two fish-bearing streams affected by the logging, and even after documented complaints, there was no action taken against Island Timberlands by the Managed Forest Council. A number of communities on Vancouver Island, the Gulf Islands, and the Interior of BC, are facing similar battles with the same experience: failure to protect their watersheds and wilderness corridors.

Down traces the origins of the problem to the crash in the markets when Gordon Campbell was premier. Down said, "In desperation, he encouraged union busting by declaring if a place was shut down for over a year, it could start back up union free; he deregulated the industry, especially the PMFL, giving them the power to basically do whatever they wanted or needed to stay alive, keep going, and be competitive. It did work, but the cost was an easier log export to markets that had cheap labour and could afford to pay more for the logs than local mills.

"We had built forestry into a long-term, self-sustainable industry that understood that the better logging practices you used and the more time/money you put into reforestation (silviculture), the better quality and faster turnover the wood was. It was a win/win for everyone and everything. It was long-term thinking! The industry and everyone in it was listening, the log it/burn it/pave it crowd were becoming dinosaurs, silenced, their rants called out as the absurd self-destructive industry threat they were. It was an awakening. We knew it was profitable through production of a higher quality product, to be environmentally friendly. Every dollar put into reforestation was money in the bank, it added value."

Down sees the problems in forestry as economic development issues, not just environmental. "The toll on communities has been harsh. Roads once busy with crew trucks in the early morning, with stores specifically open on those routes to cater to the morning rush, remain mostly empty... if Christy Clark wanted to do something in the manufacturing and reforestation sector, there is work to be done there."

Safety in the changing forest sector is an issue, according to Down, whose job included assessing danger trees. "The problems have now become more complex than just 'creating more jobs' in the industry. We now have an unbridgeable gap between seasoned veteran workers and greenhorns (newbies). We have lost knowledge and experience that was not passed on to retirement. Industry knowledge that has been passed down for over a hundred years is disappearing. This is a direct result of an absent federal government and an inept provincial government that has taken an industry for granted for a generation, ignoring the warnings from within the industry. This will, once again be paid for with lives."

Forestry is still a major industry in BC.

PREMIER CHRISTY CLARK'S government has been criticized for being too cozy with large corporations, often at the expense of "softer" topics, like the arts. Is this a legitimate complaint? I spoke with two leaders in the BC music industry to try to obtain some perspective. The first, Ed Henderson, is a composer, producer, arranger, conductor, musical director, singer, Chilliwack band member, and very talented guitar player who also serves as a member of the board of directors on the Guild of Canadian Film Composers, and a member of American Federation of Musicians and ACTRA.

Henderson said, "BC has a vibrant creative community that needs recognition and support from all sides but government can lead in this matter. The lack of such support forces young creative people to leave the province to find fulfilling work ... The lack of recording opportunities and live performance spaces in the urban centres force us to leave the province."

Henderson thinks that the problem is fairly recent, and although there may be a move back with recent funding announcements, he said, "In 2014, less than $100,000 was earned by musicians recording in film (Vancouver Musicians Association data), which cannot be blamed on government but on industry controls from the US-based American Federation of Musicians (AFM). However, policies could change so that bodies like the AFM would not have such control. Ultimately, we suffer due to US protectionism."

He thinks there are three specific actions that the BC government could take to help musicians in BC. First, "The BC government could demand that the tax credit supported film industry in BC use BC-based composers and musicians in order to receive their tax credits. Otherwise the work goes overseas to England, Bratislava or to the US."

There is also a problem just finding a location. "For musicians performing live there is a lack of performance spaces. The BC government could encourage, or force perhaps, developers to include secure live performance spaces, or all-purpose arts spaces in their developments. Vancouver has the Dance Centre, an amazing resource of rehearsal halls, performance and office space for that industry—perhaps there could be the music centre with rehearsal, performance and office space."

There are many public venues that sit empty after about 5:00 p.m. on weekdays and all weekend, and shared-space agreements would generate a better bottom line for government. Henderson feels we do not tend to value the work of performing artists, "Generally, in Canada, there is a lack of recognition of the work and contributions to society that musicians, and all artists, make. Many live as paupers in order to serve their art—that is an old story we all know. However, in places like Buenos Aires, and all over Europe, musicians are celebrated by their people, governments and businesses ... Such support and recognition, in turn, helps this music become international, but this music was celebrated and supported first at home."

Could the BC music industry create jobs or economic development? "The opportunities are enormous and well documented. Every dollar invested in the arts pays back in taxation and spinoffs between 145 percent and 300 percent . . . these investments are environmentally safe."

Henderson believes that putting money into music goes beyond simple economics. "The benefits of investment in childhood musical education are also well documented and substantial. See the value of Sara McLachlan's school and El Sistema in Caracas, Venezuela, where crime rates in the favelas decreased as much as 95 percent."

As to the future, "I am always optimistic about the future of music in BC, but I see there is much to be done to support growth in this industry."

Bob D'Eith has been an entertainment lawyer for twenty-five years; he is a musician and recording artist as part of the band Mythos; wrote the popular college-approved music business book, *A Career in Music: The Other 12 Step Program*; and served as CEO and executive director of Music BC for fourteen years. What does he think are the most important provincial issues facing musicians in BC? "The global music industry has gone through massive changes over the past decade due primarily to the impact of digitization and the Internet . . . Music piracy has resulted in a reduction of annual revenues from \$30B to \$14B."

The loss of revenue from sales means less liquidity for large music companies. "Major labels have less money to develop artists. This means that more than ever, artists are having to develop themselves. It is not enough to have talent today. Artists have to be ready to commit to working on the business side of the industry. BC artists need the tools to create careers in music. Unfortunately, for the most part colleges and universities do not prepare musicians for the new business model."

D'Eith's business model sees the BC government as a beneficiary of a thriving music industry. "The BC government can really help musicians by recognizing the role that the government can play in artist and industry development. Artists and industry need funding for music-business education, recording, marketing, promotion, showcase and tour support and other development needs.

"The return on investment can be incredible. Carly Rae Jepson recorded *Call Me Maybe* through her label 604 Records with funding from the Foundation to Assist Canadian Talent on Recordings. This song became the biggest song in the world in 2012 and launched her career. A relatively small investment resulted in a huge success story for BC."

What is BC's potential? "We have the talent, the facilities and the personnel to be a global music centre. The industry will continue with or without government support; however, there is a real danger of losing our flagship studios, key producers, artists and international business to provinces with provincial support. The BC music industry is an amazing asset for BC, and working together with the BC government, we can truly make it a global success story."

Premier Christy Clark must have agreed, because on February 11, 2016, she made a major announcement of a grant and the creation of the BC Music Fund, with some well-known BC musicians in attendance:

> Michael Bublé says British Columbia has the potential to become the "Nashville of Canada," and a new $15-million government grant will help foster a thriving local music industry.
>
> "If we can become the L.A. of the world for film, why can't we become the Nashville of the world for music?" (Premier Clark) asked.[3]

The announcement came in response to a BC report from Music Canada that was released at the same conference. It focused on rebuilding the province's music economy by creating more opportunity for up-and-coming musicians, increasing childhood music education, and allocating funds to support live music venues, among other things.

SO THERE IS work to be done, and an election in May 2017, and a track record to consider. Some argue that Premier Clark has been weak on policy. However, given her BC Jobs Plan, LNG industry development, five conditions for Enbridge, rewriting the Water Act, opening up liquor laws, housing rental subsidy, Aboriginal education outcomes, leading the country on carbon tax and climate change, establishing the Great Bear

Rainforest, local food initiatives, four balanced budgets, and investments in skills raining, plus other policies, it can be said her government has accomplished much.

The comments against Premier Christy Clark often have an element of sexism or misogyny to them. At the end of December 2015, a Facebook page run by NDP supporters since 2011 for "Crusty Clark" was finally taken down. This site had incited considerable hatred and contempt for the premier, with many demeaning images and comments, including rude poses with Premier Clark's face superimposed.

"In the social-media world," says Moe Sihota, "it is very easy to be anonymous and angry."

Sihota is a class act, a good communicator, and a lawyer. He argues his case for his party, insistent that Clark is vulnerable on the policy front, "There are two aspects to politics: there is the communications and marketing side of it, which frankly has been my side of it, I won't deny it, but there is also the aspect that brings about policy change, and on big files like climate change, children in care, affordability in housing, that impact in their daily lives, there's been very little progress."

Premier Christy Clark's integrity has many times been called into question by those who claim she has sold out her Liberal values to BC's right wing. Mark Marissen, a passionately dedicated federal Liberal, offers his perspective of Premier Christy Clark's political choices.

"She gets heavily criticized by Liberals in BC for her policies, but you have to be pragmatic in BC provincial politics. It's not like you're going to get New Democrats supporting you. Historically, Liberals are a small force in BC, so you have to build support from other political loyalties, and the last people who are going to help you are New Democrats."

According to Marissen, the NDP are simply not as progressive in BC. "They are more union based than Green. New Democrats in British Columbia are not like New Democrats in any other part of the country. I think they are solely focused on gaining power, and that has made them cynical and opportunistic."

He cites the position they took on the carbon tax in 2009. "They didn't show any kind of forward-thinking principle. They campaigned

very loudly against it, they organized, they signed people up ... and ironically, that helped Gordon Campbell win because a lot of progressive
voters supported the carbon tax, so they ended up voting for Campbell."

Sihota predicts Christy Clark's demise. "She is a formidable political
opponent ... personable and understands her political base [however] I
think everyone has a shelf life, and it comes with politics. I think the NDP
will win the next election.

"Some of the arrogance and superficiality that we have seen in the
Liberals in this administration will catch up to them. They will underestimate John Horgan's political skills. He is savvy and a remarkably
effective communicator. The [2013] election result pushed the party
back to thinking more about the economic side of the equation, in terms
of public policy. John Horgan is more focused on dealing with blue-
collar voters."

Tom Sigurdson, one of BC's leaders in advocating for good jobs for
workers, says, "It will be a different campaign in 2017. There will be two
people who relate similarly to the electorate. *Both* have great smiles." He
laughs as he reflects on this comment. "It matters. Adrian was very serious. It was hard to find a picture of him smiling; he was always focused.
The premier is also focused, but she's also always smiling."

Vaughn Palmer has seen a number of party leaders rise and fall
during his tenure in the press gallery. "Horgan [is] a stronger communicator than the last ex-staffer to take the job [Dix]. One of the interesting
things to watch in the legislature is the premier working to get his number. She did the same with Dix, with results that are a matter of record."

Mike McDonald says Clark's track record will stand out, "Christy has
been underestimated as someone having firm, frugal principles. Today,
the government is on track to wipe out operating debt by 2019–20, which
hasn't happened since the mid-1970s. BC is the only province in Canada running a surplus, and it's run four straight under Christy. No one
thought she could do it."

According to Laura Miller, within the Liberal Party, "There is deep
admiration for her in terms of her dedication, her contribution, and the
sacrifices she has made, particularly in terms of her time with Hamish,

to lead the party. [Members] are excited about the direction in which she is leading the government. She is happy to support the members of the party; the members and their views are so important to her. They feel valued."

When asked about the people who are so negative toward the premier, Miller replied, "At the end of the day, if the average person could sit down and share a coffee or a beer with any politician, they would see them all very differently as regular people."

Don Guy said of Clark, "She won a ton of respect from people across Canada... When I hear people talk about her now, it is with respect and admiration. They know she is a fighter, they know it is a mistake to underestimate her. They know she is willing to fight for what she believes in. You hear this wherever you go in BC, and wherever you go in Canada, whether Calgary, Toronto, or Ottawa."

Former party president Floyd Sully, a staunch Liberal, has considerable admiration for Christy Clark and speaks with passion when he talks about her accomplishments, "Christy Clark got where she is through her own hard efforts, fought her way through the obstacles. She did it all on her own; no one opened a single door for her. No one handed her the leadership the way they did with Campbell.

"She wasn't part of any old boy's club."

Sharon White, current party president, who comes from the Conservative part of the government coalition, said, "I see someone who is smart, passionate, kind, caring, and funny, and has a great love for life. I see someone who is a friend, with loads of personality; that is who I know. I think that the public perception is not the same; probably that she is more opportunistic, and that she is not genuine. They may see her as too aggressive. I see a woman working hard at a demanding job. I don't consider that aggressive if she's doing the job as it is required."

When asked about the gap between how people see the premier when they work with her, and how they see her in regular and social media, White said, "I think that we live in a polarized province, so people who are on the other side of the political spectrum from us will gravitate toward the most negative perception of the premier... She's a magnet for all of the negative things going on."

Former Liberal Leader Gordon Wilson is very supportive of the current Liberal Leader and observed, "Christy Clark is an attractive woman. A lot of female leaders think that they have to act and dress like men. If some man had said we are going to close the bridge to do yoga, do you think we would have had Photoshopped images of a man doing yoga? A lot of people attack her for being a woman, and the most vicious attacks come from other women."

When political opponents attack any issue, the glass is always half-full, and the contents are suspicious. This extremely negative conversation has led many BC residents to simply tune out. Democracy only functions well when we are all tuned in, and I hope this book offers readers a chance to engage on some important issues.

A final word, from Premier Christy Clark, on governing BC and the political journey that led her to where she is today:

"My dad was a guy who liked to have fun with politics. For him, being a Liberal set him apart as an iconoclast, which he enjoyed. He was also a member of the teachers' union and wanted to feel comfortable as a free-enterprise voter when he was not a New Democrat."

Her dad's commitment to the Liberal Party was, according to her, part of his beliefs, but not part of his ambition.

"I am about trying to get things done, I want to be in government and find a way to change the world through politics. My dad was more interested in conversations around the dining room table."

I asked her about those who claim that the party she inherited from Gordon Campbell is little more than the Social Credit coalition under a different name.

"Yes, it is important that the BC Liberal Party is a coalition, but we are a totally different party. In our party, there are left-wing BC Liberals and right-wing BC Liberals; there are green (pro-environment) Liberals and brown (pro-resource job) Liberals."

She also points out that British Columbia has changed dramatically in the past decades, especially the labour force. "Look how the world has changed: 20 percent of the workforce is union; it used to be 80 percent. Small business is a much bigger player now."

For Clark the party is less about rigid ideology and more about finding ideas that work. "Basically, these are community leaders who have come together. Half the caucus I represent is brand new and has never run for office before. They ran because they believed in what we were doing at a time when it was the darkest time."

As a leader, she sees her role as being able to lead collaboration.

"Our party is a group of people driven by the communities they come from. We are broadly organized around the idea that British Columbia's economy has to grow. We have to find ways to say yes to growth. Our Five Conditions are the perfect example, setting out a path to get to yes. This is very different from trying to find ways to say no. You've got to meet environmental standards; First Nations have to be part of the process; you have to respect communities along the way; there has to be a business case so that everyone benefits from the project. It has to make sense.

"We are not ideologues who are pro-union or pro-business. Our party includes feminists, workers, environmentalists, retirees, business owners, resource people, business people, and families."

In March 2016, Premier Christy Clark became Canada's longest-serving female first minister. Her big brother Bruce knows she is up for the challenge of governing and contesting the next election as leader.

"It's a hell of a big job, [as] but people have learned, don't underestimate her."

Her smile has become almost a trademark, and I ask her if she's happy.

"I have the most challenging and fulfilling job I could imagine. I have a son I love. I have a lot of incredibly supportive friends. I have my health. Happiness is a product of how you feel about yourself. It is the product of living a fulfilling life. It is not an end in itself. I don't strive to be happy. I strive to be fulfilled."

NOTES

CH. 1

1 The quoted epigraphs that appear at the start of this and all following chapters are excerpted from David Brooks, *The Road To Character*, New York: Random House, 2015.

2 Stephen Hui, "NDP leader John Horgan skewers Christy Clark on 'offensive' Burrard Bridge yoga closure," *Georgia Straight*, June 11, 2015.

3 "Burrard Bridge yoga plan mocked by singer Raffi Cavoukian on Twitter," *CBC News*, June 9, 2015.

4 Emily Jackson, "Only in B.C.: Raffi calls out premier over Burrard Bridge yoga event," *Metro News*, June 7, 2015.

5 "Gregor Robertson won't attend 'Om the Bridge' as yoga backlash grows," *CBC News*, June 11, 2015.

6 "Premier's Statement on the International Day of Yoga," *BC Gov News*, June 12, 2015.

7 Interview with Chip Wilson, November 14, 2015 (by telephone).

CH. 2

1 Wikipedia, British Columbia Social Credit Party, entry cited October 15, 2015.

2   *The Encyclopedia of British Columbia*, knowbc.com, Harbour Publishing,
    Madeira Park, BC, October 15, 2015.

3   Interview with Mike McDonald, September 17, 2015 (by telephone).

4   Conversation with Mark Devereux, October 8, 2015, after review of Mike
    McDonald's University Model Parliament records.

3   Interview with Bruce Clark, July 9, 2015.

CH. 3

1   Interview with Christy Clark, October 8, 2015.

2   Interview with Bruce Clark, July 9, 2015 (by telephone).

3   Interview with Dawn Clarke, October 17, 2015 (by telephone).

4   Interview with Keith Bradley, July 10, 2015.

5   Interview with Gordon Wilson, July 18, 2015.

CH. 4

1   Interview with Mark Marrisen, July 8, 2015.

2   Interview with Derek Raymaker, July 14, 2015 (by telephone).

3   Cecil Favron, "A Peak Back in Time: The 1989 Christygate Scandal,"
    Peak (SFU), April 7, 2015.

4   Interview with Laura Miller, September 14, 2015 (by telephone).

CH. 5

1   Interview with Dawn Clarke, October 17, 2015 (by telephone).

2   Interview with Athana Mentzelopoulos, August 29, 2015 (by telephone).

3   en.wikipedia.org/wiki/Elijah_Harper, October 15, 2015.

CH. 6

1   Interview with Gordon Wilson, July 18, 2015.

2   Interview with Floyd Sully, October 16, 2015 (via email).

3   Interview with Gordon Wilson, October 18, 2015.

4   "Fantasy Gardens and Bill Vander Zalm, Following is a chronology of
    the troubles Fantasy Gardens brought Bill Vander Zalm," *Vancouver Sun*,
    September 13, 1991.

5   Patricia Lush, "Harcourt battling B.C. businesses' fear of NDP," *Globe and
    Mail*, April 20, 1991.

6   Interview with Floyd Sully, July 17, 2015.

7   "Wilson in debate might help Socreds, hurt N D P," *Province,* October 3, 1991.

8   Interview with Floyd Sully, October 16, 2015 (via email).

9   Brian Kieran, "Wilson's skewed format for T V debate tonight," *Province,* October 8, 1991.

10 Vaughn Palmer, "Johnston lets slip the mask of optimism," *Vancouver Sun,* October 7, 1991.

11 Vaughn Palmer, "Win or lose: one moment in a debate," *Vancouver Sun,* October 8, 1991.

12 Robert Matas and Deborah Wilson, "Johnston gets tough in B.C. debate, tells N D P, 'Come clean'," *Globe and Mail,* October 9, 1991. Canadian Press, "Johnston attacks N D P policies in B.C. debate," October 9, 1991. "Premier dominates debate with attack on Harcourt," *Ottawa Citizen,* October 9, 1991.

13 Tom Barrett, "Grits surge into 2nd in B.C. race; debate showing propels Liberals past Socreds," *Vancouver Sun,* October 12, 1991.

14 Robert Matas, "Surge by Liberals shakes up B.C. campaign," *Globe and Mail,* October 16, 1991.

15 Keith Baldrey, "The Upside Down Campaign: N D P bobbed, Socreds sank, then along came Wilson," *Vancouver Sun,* October 18, 1991.

CH. 7

1   Email response from Linda Reid, August 7, 2015.

2   Ian Austin, "Cowie broadsides his boss: Liberal dissidents keeping the heat on leader Wilson," *The Province,* October 11, 1992.

3   Interview with Gordon Wilson, July 18, 2015.

4   Interview with Vaughn Palmer, July 10, 2015 (email and telephone).

5   Interview with Keith Baldrey, July 10, 2015 (email and telephone).

CH. 8

1   Jane Taber, "How I spent my summer vacation; Young Liberals learned to pack a hall; some young Tories threw up in the halls," *Montreal Gazette,* September 2, 1992.

2   Interview with Don Guy, September 15, 2015 (by telephone).

CH. 9

1   Interview with Sharon White, September 14, 2015 (email and telephone).

2   Interview with Derek Raymaker, July 14, 2015 (by telephone).

3   Vaughn Palmer, "'No Girls Allowed' woes for the Liberals," *Vancouver Sun*, May 31, 1995.

CH. 10

1   Senate.gov/?page_id=81, August 29, 2015.

CH. 11

1   Les Leyne, "Clark woos voters with tax-cut lure, NDP premier bashes Liberal leader for vow to slash spending," Victoria *Times Colonist*, April 24, 1996.

2   Don Hauka, "Campaign trail strewn with goodies (not huge) for regular folks," *Province*, April 24, 1996, A5.

3   Mike Smyth, "Just folks' image bombs on Campbell, Liberals still the party to beat despite NDP resurgence," Canadian Press, March 20, 1996.

4   Jim Hume, "Talk Politics, Politics comes down to shirts," Victoria *Times Colonist*, March 23, 1996.

5   Geoffrey Castle, "Election 96 – Quotes," Victoria *Times Colonist*, May 29, 1996.

6   Les Leyne and Dirk Meissner, "Socred leader quits, urges B.C. to vote right; Despite secret meetings with Liberals, Larry Gillanders says there was no deal for him to step aside," Victoria *Times Colonist*, May 25, 1996.

7   Vaughn Palmer, "B.C. Liberals resort to old-style shenanigans: Liberals blow chance to breathe fresh air into politics," *Vancouver Sun*, May 27, 1996.

8   Statement of Votes, 36th provincial election, May 28, 1996, Elections BC.

9   Rafe Mair, "B.C.'S non-socialists should have taken on clark together: Lack of a pact cost the Liberals the election," *Financial Post*, May 31, 1996.

10 "Reform MLA quits party," *Globe and Mail*, November 29, 1997.

11 Don Hauka, "NDP reneging: Liberal," *Province*, July 16, 1996.

12 Jim Beatty, "Forestry firm that felled trees in park warned of tough new fines: The Liberal environment critic says Interfor should not have been allowed to keep the timber harvested from Garibaldi," *Vancouver Sun*, July 31, 1996.

13 Scott Simpson, "Firing up Burrard plant will increase air pollution, Liberal MLA says," *Vancouver Sun*, July 31, 1996.

14 Mike Smyth, "Upstart goads the pit bull: Christy Clark emerges as the Liberal's most talented critic, easily nailing elusive targets," *Province*, May 28, 1999.

15 Interview with Moe Sihota, October 12, 2015 (by telephone).

16 Ibid.

CH. 12

1 Mike Smyth, "Liberals in labour: Christy Clark isn't the only one who's due; Gordon Campbell's got a lot to deliver, too," *Province*, August 5, 2001.

2 Lori Culbert, "Liberals not shying away from showdown with teachers: Education Minister Christy Clark vows teachers will be an essential service," *Vancouver Sun*, June 6, 2001.

3 Edward Greenspon, "A most powerful baby," Inside Politics, *Globe and Mail*, August 11, 2001.

4 Janet Steffenhagen and Judith Lavoie, "Clark favours education ratings," Victoria *Times Colonist*, September 26, 2001.

5 "Parents allowed in schools, legislation declares," *Kamloops Daily News*, August 3, 2001.

6 Lori Culbert, "Pregnant MLA in name game debate," *Calgary Herald*, August 13, 2001.

7 Interview with Guy Monty, October 7, 2015 (by telephone).

8 Interview with Jatinder Rai, September 15, 2015 (by telephone).

9 Interview with Don Doyle, October 6, 2015.

10 Jim Beatty, "Campbell dramatically remakes B.C. cabinet," *Prince Rupert Daily News*, January 27, 2004.

11 Vaughn Palmer, "Clark's surprise for Campbell hinges on political ambition, after all," *Vancouver Sun*, September 17, 2004.

12 Michael Smyth, "There's more to the Clark resignation than she's telling us," *Province*, September 17, 2004.

CH. 13

1 Jim Beatty with files from Lori Culbert and Petti Fong, "Politicians scrambling to deal with raids' fallout," *Vancouver Sun*, December 30, 2003.

2 Interview with Mark Marissen, January 31, 2016.

3 Petti Fong, "Marissen a top Martin organizer," *Vancouver Sun*,
December 30, 2003.

4 Interviews with Mark Marissen, July 8, 2015, and January 31, 2016.

5 CBC News, British Columbia news, April 10, 2013.

6 *An Audit of Special Indemnities*, Office of the Auditor General of British
Columbia, December 2013, page 5.

7 Ibid, page 7.

8 Harvey Oberfeld, "Full BC Rail Story NOT Yet Told," *Keeping it Real…*
(blog), January 9, 2014. harveyoberfeld.ca/blog/full-bc-rail-story-
not-yet-told/.

CH. 14

1 "B.C. deputy premier gives birth: a first in province's history," Canadian
Press NewsWire, August 27, 2001.

2 Interview with Mark Marissen, July 8, 2015.

3 Hansard, BC Legislative Assembly, March 7, 2005.

4 Interview with Derek Raymaker, July 14, 2015 (by telephone).

CH. 15

1 Interview with Mark Marissen, January 31, 2016.

2 Interview with Christy Clark, October 8, 2015.

3 Marcie Good, "Christy Clark, Nothing Better than Politics," *BC Business*,
August 1, 2005.

CH. 16

1 "Women's Suffrage," *Canadian Encyclopedia* (online edition), December 4,
2015.

2 Jonathan Fowlie, "Clark marks first day in legislature as premier; New MLA
for Vancouver-Point Grey touts industry's bid to secure shipbuilding deal,"
Vancouver Sun. May 31, 2011.

3 Interview with Rich Coleman, December 14, 2015 (by telephone).

4 Tiffany Crawford, "John van Dongen quits BC Conservatives; John
Cummins holds on to leadership," *Vancouver Sun* (online edition),
September 23, 2012.

5 Vaughn Palmer, "A persistent John van Dongen launches another instalment in the BC Rail saga," *Vancouver Sun*, November 8, 2012.

6 Interview With Josie Tyabji, December 4, 2015 (via email).

CH. 17

1 "Natural Gas: A Primer," Government of Canada, Natural Resources Canada, nrcan.gc.ca/energy/natural-gas/5641.

2 Interview with Rich Coleman, December 14, 2015 (by telephone).

3 Interview with Jas Johal, September 7, 2015 (by telephone).

4 "BC seeks 'fair share' in new Gateway pipeline deal, Province lays out 5 criteria for provincial approval of all new crude pipelines," *CBC News* (online edition), July 24, 2012.

5 Interview with Christy Clark November 25, 2015.

CH. 18

1 Cambridge Advanced Learner's Dictionary & Thesaurus, Cambridge University Press, online edition, cambridge.org, December 6, 2015.

2 Interview with Guy Monty, October 7, 2015 (by telephone).

3 Jonathan Fowlie, "Christy Clark takes on Richard Branson over 'naked' blog post," *Vancouver Sun* (online edition), May 29, 2012.

4 Rob Shaw, "DJ who asked premier her thoughts on being a "MILF" is fired by Courtenay radio station," Victoria *Times Colonist* (online edition), January 10, 2013.

CH. 19

1 BC Business Council, "A Decade by Decade Review of British Columbia's Economic Performance," November 5, 2012 (online edition) bcbc.com.

2 Interview with Mark Marissen, December 2, 2015.

3 Jas Johal, "Gordon Wilson urges British Columbians to 'come home' and support the BC Liberals," *Global* BC *News* (online edition), May 5, 2013.

4 Ian Bailey, "Gordon Wilson rejoins Liberal camp, backing Clark ahead of B.C. election," *Globe and Mail* (online edition), May 5, 2013.

5 "Gordon Wilson flip flops again, this time from no allegiances to signing on with Christy Clark," CBC Radio, *Early Edition*, May 6, 2013.

6   Gordon Wilson, "Here's why Gordon Wilson supports Christy Clark," *Vancouver Sun*, May 8, 2013.

7   Jas Johal, "Movement underway within BC Liberals to oust Christy Clark; Dix has no comment," *Global News*, May 9, 2013.

CH. 20

1   Andy Radia, "Report into B.C.'s ethnic outreach scandal indicates misuse of government resources," *Canada Politics, Yahoo News*, March 14, 2013.

2   Ian Bailey, "Liberal ethnic-outreach scandal claims key player," *Globe and Mail*, March 20, 2013.

3   Interview with Don Guy, November 8, 2015 (by telephone).

4   Interview with Laura Miller, September 14, 2015 (by telephone).

CH. 21

1   Jonathan Fowlie, "Christy Clark Politician first; B.C.'s premier has been educated in the hard-knocks school of politics for almost her entire life," *Vancouver Sun*, April 27, 2013.

2   Stephen Smart, "B.C. election debate turns spotlight on leadership styles, Christy Clark and Adrian Dix have led very different campaigns," *CBC News*, April 29, 2013.

3   Ian Bailey and Justine Hunter, "B.C. leaders trade punches in election's only televised debate," *Globe and Mail*, April 30, 2013.

4   "New Democrat John Horgan says Liberals on smear campaign over Kinder Morgan," Canadian Press, *Vancouver Sun* (online edition), May 7, 2013.

5   Gordon Hoekstra, "Dix's pipeline flip flop key factor in election outcome: union," *Vancouver Sun* (online edition), May 18, 2013.

6   Interview with Tom Sigurdson, January 15, 2016.

CH. 22

1   Interview with Premier Clark, December 8, 2015.

2   Justine Hunter, "MLA Ben Stewart steps aside so Clark can run in by-election," *Globe and Mail*, June 5, 2015.

3   Jonathan Fowlie, "Ben Stewart steps aside in Kelowna for B.C. Premier Christy Clark," *Vancouver Sun*, June 6, 2015.

CH. 23

1 Union of BC Indian Chiefs, online source, February 24, 2016. ubcic.bc.ca/

2 First Nations Summit, online source, February 24, 2016. fns.bc.ca/about/
about.htm.

3 BC Assembly of First Nations, online source, February 24, 2016. bcafn.ca/
history/.

4 Nuu-chah-nulth Tribal Council, online source, February 15, 2016.
nuuchahnulth.org/tribal-council/welcome.html.

5 Indigenous and Northern Affairs Canada, Aboriginal peoples and communities, First Nations in Canada, February 11, 2016, aadnc-aandc.gc.ca/eng/1
307460755710/1307460872523.

6 Justine Hunter, "Breaking through the First Nations wall," *Globe and Mail*,
July 14, 2013.

7 Larry Pynn, "First Nations' economic clout the result of decades of court
decisions," *Vancouver Sun*, May 30, 2015

8 Justine Hunter and Dianne Rinehart, "Campbell vows referendum for
proposed Nisga'a treaty: B.C. Liberal leader says voters should have opportunity to approve or kill deal," *Vancouver Sun*, July 25, 1998.

9 Dianne Rinehart, "Close vote on Nisga'a deal 'disappointing': Premier
Glen Clark and provincial Aboriginal Affairs Minister Dale Lovick stand
united in lamenting early poll results reporting only 51-per-cent of eligible
Nisga'a voters support B.C.'s first modern-day treaty," *Vancouver Sun*,
November 12, 1998.

10 "Right treaty, wrong process: Using closure to make the Nisga'a treaty an
irrevocable law is an abuse of democracy. The Clark government's action is
not acceptable." *Vancouver Sun*, April 17, 1999.

11 "Government taints treaty by trampling democracy," *Vancouver Sun*, April
21, 1999.

12 Mark Hume, "National Post said NDP cuts off debate, passes Nisga'a treaty;
Ruling NDP shuts down debate on Nisga'a treaty: Liberals accuse embattled
premier of despotism," *National Post*, April 23, 1999.

13 Barbara McLintock and Ian Austin, "House passes Nisga'a treaty: Liberals
walk out of legislature in protest after NDP imposes closure on marathon
debate," *Province*, April 23, 1999.

14 *Honouring the Truth, Reconciling for the Future, Summary of the Final Report of the Truth and Reconciliation commission of Canada,* online edition, May 31, 2015. trc.ca/websites/trcinstitution/File/2015/Findings/Exec_Summary_2015_05_31_web_o.pdf.

15 *Truth and Reconciliation Commission of Canada,* online, February 12, 2016. trc.ca/websites/trcinstitution/index.php?p=4.

16 "Calls to Action," *Truth and Reconciliation Commission of Canada.* trc.ca/websites/trcinstitution/File/2015/Findings/Calls_to_Action_English2.pdf.

17 "Is BC Getting it Right? Moving toward Aboriginal education success in British Columbia," Canadian Education Association. cea-ace.ca/education-canada/article/bc-getting-it-right.

18 Tracy Sherlock, "B.C. outperforms other provinces on aboriginal graduation rates, report says," *Vancouver Sun,* May 1, 2014.

19 Geordon Omand, "Aboriginal high school graduation rate at record high in B.C.," Canadian Press reported by CBC, December 29, 2015. cbc.ca/news/canada/british-columbia/aboriginal-graduation-rate-1.3383908.

20 Tom Fletcher, "Aboriginal families prepare for inquiry," *Merritt Herald,* February 3, 2016. merrittherald.com/aboriginal-families-prepare-for-inquiry/.

21 Lori Culbert, "B.C. hires high-profile native leader Ed John to improve adoptions for aboriginal kids," *Vancouver Sun,* September 9, 2015.

22 Lindsay Kines, "Premier: George Abbott out because B.C. treaty process needs reform," Victoria *Times Colonist,* March 26, 2015. timescolonist.com/news/local/premier-george-abbott-out-because-b-c-treaty-process-needs-reform-1.1804829#sthash.d7BltDTL.dpuf.

23 Ian Bailey, "Veteran politician George Abbott hits the books," *Globe and Mail,* October 17, 2013.

24 *Andrew MacLeod,* "George Abbott Quits Liberals He Sought to Lead, 'It was no single thing. I've thought about it for a while,'" *Tyee,* June 30, 2015. thetyee.ca/News/2015/06/30/George-Abbott-Quits-Liberals/.

25 Robin M. Junger, Joan M. Young, & Brent Ryan, *"Supreme Court declares Aboriginal title in Tsilhqot'in Nation v. British Columbia,"* McMillan LLP June 2014. mcmillan.ca/mobile/showpublication.aspx?show=108648.

26 Vaughn Palmer, "No quick fix for gulf between province and First Nations, Both acknowledge lack of a 'common understanding' of their relationship," *Vancouver Sun*, September 18, 2015.

27 Westbank First Nation, Government, online source, February 28. wfn.ca/government.htm.

28 Justine Hunter, "Final agreement reached to protect B.C.'s Great Bear Rainforest," *Globe and Mail*, February 1, 2016.

29 Ibid.

30 Interview with Christy Clark, February 18, 2016.

CH. 24

1 Interview with Christy Clark.

2 "Premier releases Canada Starts Here: The BC Jobs Plan," Office of the Premier, September 22, 2011. news.gov.bc.ca/stories/premier-releases-canada-starts-here-the-bc-jobs-plan.

3 Iglika Ivanova, "Two years in, the BC Jobs Plan failing to deliver, Opinion: Positions created have been primarily temporary, haven't kept up with growing population," *Vancouver Sun,* January 9, 2014.

4 Les Leyne, "B.C. job stats not great, but not bad," Victoria *Times Colonist*, January 9, 2016.

5 Mike Clarke, "Site C dam: How we got here and what you need to know, Dam first proposed in the 1950s as the third in a series of four dams, two of which were built," *CBC News*, December 16, 2014.

6 Interview with Christy Clark.

7 Mike Clarke, "Site C dam: How we got here . . .," *CBC News*, December 16, 2014.

8 Justine Hunter, "B.C. premier plays wedge politics with Site C dam," *Globe and Mail*, October 4, 2015.

9 "B.C. launches Skills for Jobs Blueprint to re-engineer education and training," Ministry of Jobs, Tourism and Skills Training, April 14, 2014.

10 "B.C. to refocus education, training plan to fill 1 M jobs, Ministers announce 'blueprint' to better match education resource allocation to B.C. labour market," Canadian Press, April 30, 2014.

11 Justine Hunter, "B.C. to boost skills training programs in public education," *Globe and Mail*, April 29, 2014.

12 Jock Finlayson, "Skill Shortages: Weighing Employers' Views," British Columbia Business Council, April 9, 2014. bcbc.com/bcbc-blog/2014/skill-shortages-weighing-employers-views.

13 Vaughn Palmer, "On LNG skills training, labour works the apprenticeship angle, Christy Clark makes a place for unions after 10 years of the Gordon Campbell isolation chamber," *Vancouver Sun*, April 9, 2014.

14 Wayne Moore, "BC Fed: Workers ignored," *Castanet*, December 26, 2015. castanet.net/news/BC/154695/BC-Fed-Workers-ignored.

15 Les Leyne, "Clark shows the energy of a winner," Les Leyne, Victoria *Times Colonist*, September 21, 2013.

16 Interview with Tom Sigurdson, January 15, 2016.

17 Interview with Christy Clark.

18 Les Leyne, "Clark shows the energy of a winner," Victoria *Times Colonist*, September 21, 2013.

19 Robert Barron, "Premier pushes benefits of B.C.'s natural gas supply during Nanaimo stop—Clark also announces $15M for Malahat safety," *Nanaimo Daily News*, October 31, 2013

20 Ibid.

21 Vaughn Palmer, "John Horgan holds his nose on LNG," *Vancouver Sun*, June 16, 2015.

22 Roger Gibbins, "Like her or loathe her, B.C. Premier Christy Clark is possibly Canada's most influential energy player, *Alberta Oil*, February 03, 2014.

CH. 25

1 Vaughn Palmer, "Wilson an LNG skeptic as recently as April," *Vancouver Sun*, October 31, 2015.

2 Les Leyne, "Wilson's appointment a bit mystifying," Victoria *Times Colonist*, November 5, 2013.

3 Rafe Mair, "Gordon Wilson finds religion on LNG... for $12,500 a month," *Common Sense Canadian*, November 4, 2013. commonsensecanadian.ca/gordon-wilson-fracks-legacy-becoming-LNG-booster/.

4 Brent Stafford, "Premier wouldn't appoint unqualified people to vital trade posts," The Duel, *24 Hours Vancouver*, November 3, 2013. vancouver. 24hrs.ca/2013/11/03 premier-wouldnt-appoint-unqualified-people-to-vital-trade-posts.

5 Interview with Gordon Wilson.

CH. 26

1 Colleen Kimmett, "New West to revive farmers' market coupon program," *Tyee*, February 25, 2011.

2 Nutrition Coupon Program, BC *Association of Farmers' Markets*, January 16, 2016. bcfarmersmarkets.org.

3 "$2 M supports expanded farmers' market coupon program," Ministry of Health, July 20, 2013. news.gov.bc.ca/stories/2m-supports-expanded-farmers-market-coupon-program.

4 "Mountain pine beetle (factsheet)," Natural Resources Canada, Government of Canada, January 16, 2016. nrcan.gc.ca/forests/fire-insects-disturbances/top-insects/13397.

5 Travis Lupick, "Calls for B.C. to drop LNG ambitions to preserve a leadership role on climate change," *Georgia Straight*, November 25, 2015.

6 Justine Hunter, "Changes to B.C.'s climate-change law to depend on national targets," *Globe and Mail*, December 09, 2015.

7 Tom Fletcher, "Christy Clark on greenhouse gases and government ads," *Maple Ridge and Pitt Meadows News*, December 15, 2015.

8 Darren Handschuh, "Climate change input sought," *Castanet*, Jan 16, 2016. castanet.net/news/BC/156237/Climate-change-input-sought.

9 Interview with Christy Clark.

CH. 27

1 Bill Tieleman, "Nestle Pays $2.25 to Bottle and Sell a Million Litres of BC Water, I repeat: Nestle pays $2.25 to bottle and sell a million litres of BC water," *Tyee*, February 24, 2015.

2 SumOfUs, January 16, 2016. action.sumofus.org/a/bc-bottled-water/13/2/.

3 Dan Fumano, "Outrage boils over as B.C. government plans to sell groundwater for $2.25 per million litres," *Province*, March 6, 2015.

4   "Nestlé faces renewed criticism as B.C. drought continues, Companies like Nestlé will pay $2.25 to bottle a million litres of water starting in 2016," *On the Coast,* CBC News, July 10, 2015.

5   facebook.com/judi.tyabji/posts/10153042171231964.

CH. 28

1   "Group sex, illegal pot get OK in Vancouver, but wine in grocery stores? No way," *National Post,* Brian Hutchinson, January 19, 2016. news.nationalpost.com/full-comment/brian-hutchinson-group-sex-illegal-pot-get-ok-in-vancouver-but-wine-in-grocery-stores-no-way.

2   BC Ferries, About BC Ferries, Our History, March 2, 2016. bcferries.com/about/history/history.html.

3   "Michael Bublé praises British Columbia's new music fund," Laura Kane, *Globe and Mail,* February 11, 2016. theglobeandmail.com/news/british-columbia/michael-buble-praises-british-columbias-new-music-fund/article28738074/.

LIST OF INTERVIEWS

The following is a list of interviews conducted by the author in the course of researching and writing this book. Interviews took place in person and by phone, email, or Skype. Excerpts from these interviews are quoted throughout the book, by permission of the interviewees listed below.

Kim Baird (email), February 23, 2016.

Keith Baldrey (telephone), July 19 and 20, 2015, plus follow-up email.

Brad Bennett (telephone), September 9, 2015.

Bruce Clark (telephone) July 9, 2015 and January 12, 2016, plus in-person discussion regarding photos used in the book.

Premier Christy Clark (in person) October 8, 2015; (in person) November 25, 2015; (in person) December 8, 2015; and (telephone) February 29, 2016.

Dawn Clarke (telephone), October 17, 2015, plus email exchanges.

Rich Coleman (telephone), December 14, 2015.

Bob D'Eith (email) February 4, 2016, following in-person discussion.

Jason Down (email), February 2, 2016, plus Facebook exchange.

Dan Doyle (telephone), September 27, 2015.

Jenny Garden (email), February 9, 2016.

Don Guy (email), November 7, 2015, plus email exchanges.

Ed Henderson (email) February 9, 2016, following in-person discussion.

Wendy John (email and telephone), January 14, 2016; February 6, 2016; and
    February 8, 2016.

Jas Johal (in person), September 7, 2015, plus email exchanges.

Chief Robert Louie (telephone), February 3, 2016, plus mail and email exchanges.

Lee Mackenzie (email), July 13, 2015.

Mark Marissen (telephone and email) July 8, 2015; December 2, 2015; and
    January 31, 2016.

Athana Mentzelopoulos (telephone), August 29, 2015.

Mike McDonald (telephone), September 17, plus email exchanges.

Laura Miller (telephone), September 14, 2015.

Guy Monty (telephone), October 7, 2015, plus email exchanges.

Vaughn Palmer (in person) July 10, 2015, and (email) July 13 to 17, 2015.

Grand Chief Stewart Phillip (telephone), February 3, 2016.

Jatinder Rai (telephone), September 15, 2015.

Derek Raymaker (telephone), July 14, 2015.

Linda Reid (email), August 7, 2015.

John Reynolds (telephone), September 9, 2015.

Chief Ellis Ross (telephone), February 4, 2016.

Tom Sigurdson (telephone), January 15, 2016.

Moe Sihota (telephone), October 12, 2015.

Floyd Sully (Skype), July 17, 2015, and (email) October 16, 2015.

Jean Teillet (email), February 9, 2016, following in-person discussion.

Josie Tyabji (email) December 4, 2015.

Sharon White (telephone), September 14, 2015.

Chip Wilson (telephone), November 14, 2015.

Gordon Wilson (in person), July 18, 2015, and October 18, 2015.

INDEX

ABOUT
THE AUTHOR

JUDI TYABJI is the author of *Political Affairs* (1994) and *Daggers Unsheathed: The Political Assassination of Glen Clark* (2002), two best-selling books about BC politics. She is the former MLA for Okanagan East (1991–96) and served on Powell River City Council from 1999 to 2001. She hosted a daily live radio program, *The Judi Tyabji Show*, from 1993 to 1996, and the daily live television programs *Tyabji* (CHEK TV, 1996–98) and *The Westerly* (SHAW, 2001–02). She has been a book reviewer for the *Vancouver Sun* and *Monday Magazine* and an editorialist for radio stations in Victoria and Kelowna. She is in private business, operating a sheep farm in Powell River, BC, and was the founder and president of the Pebble in the Pond Environmental Society from 2008 to 2016. She and her husband, Gordon Wilson, share five grown children and five grandchildren. Judi was born in Calcutta, India, and immigrated to Canada with her family as a child, growing up in the Okanagan Valley. She holds a political science degree from the University of Victoria.